Transformative Approaches to Career-Ready Education

Ivan Gratchev · Krishnasamy T. Selvan ·
Hugo G. Espinosa
Editors

Transformative Approaches to Career-Ready Education

Springer

Editors
Ivan Gratchev
School of Engineering and Built Environment
Griffith University
Southport, QLD, Australia

Hugo G. Espinosa
School of Engineering and Built Environment
Griffith University
Nathan, QLD, Australia

Krishnasamy T. Selvan
Department of Electronics and Communication Engineering
Sri Sivasubramaniya Nadar (SSN) College of Engineering
Kalavakkam, Tamil Nadu, India

ISBN 978-981-96-7017-8 ISBN 978-981-96-7018-5 (eBook)
https://doi.org/10.1007/978-981-96-7018-5

This Springer imprint is published by the registered company Springer Nature Singapore Pte Ltd.
The registered company address is: 152 Beach Road, #21-01/04 Gateway East, Singapore 189721, Singapore

Foreword

As the country with the largest youth population globally, India requires innovative educational reforms that can prepare the next generation for the uncertainties of the 21st century. These reforms are broadly outlined in the National Education Policy (NEP 2020) and are being actively detailed and implemented at the national and state levels. In the context of these significant developments, it is fitting that the book *Transformative Approaches to Career-Ready Education* is being published. The collaborative effort by the book's editors, Ivan Gratchev and Hugo G. Espinosa from Australia and Krishnasamy Selvan from India, is one of the first fruits of a historic agreement cemented in 2023 by India and Australia to facilitate the mutual recognition of qualifications in both countries for access to higher education.

Eminent professors and researchers from the Indian and Australian higher education systems have contributed chapters that delve into various aspects of higher education. I anticipate that this book will stimulate contemplation on improved educational policies and practices within the two countries and beyond.

A vital purpose of education should be to cultivate a happy and sustainable society where students acquire strong fundamentals, become lifelong learners, and develop as "good human beings" with compassion, empathy, courage, resilience, and solid ethical values—a recurring theme in NEP 2020. This book demonstrates how universities and teachers can broaden their educational goals to encompass these abilities and values. The book's contents are appropriately structured, covering essential aspects such as university roles, goals for higher education, leadership, effective communication, work-integrated learning, problem-solving, and active learning. All the chapters exhibit consistency with the core ideas discussed in NEP 2020.

One of the key recommendations of NEP with respect to higher education is strengthening research in the university system. The policy highlights the critical need to better manage research at all levels and research to address the specific needs of industry and social objectives. NEP recommends the creation of a National Research Foundation (NRF), whose role is to nurture a vibrant research ecosystem through adequate funding, mentoring, and support of multidisciplinary research. Australian and Indian institutions can benefit from this initiative leading to the creation of a vibrant and productive research system catering to a variety of national/international endeavours. Through NRF, collaborations between Australian and Indian institutions can be facilitated with twinning programmes and strategic partnerships for expanding and strengthening research collaborations for the two-way movement of talented research students. This book recognizes the changing dynamics within education and emphasizes the crucial need to furnish young minds with adaptable mindsets and versatile skill sets essential for confidently navigating an uncertain future and growing into stronger individuals who can contribute to the task of nation-building more effectively.

Successful, satisfying careers are the result of the long-term professional journey of individuals, which requires constant reflective engagement. This book is a step toward encouraging academic leadership, educators, and students to reflect on their hopes and expectations for higher education in the 21st century. I am sure that the book will be received with great seriousness and enthusiasm and will have wide-ranging use among a diverse set of stakeholders concerned with higher education and employment.

The Foreword author, Dr. K. Kasturirangan, recently passed away, marking a significant loss to the academic and scientific communities.

Bengaluru, India

K. Kasturirangan
Chairman, National Education Policy Committee
Chairman, National Steering Committee for National Curriculum Frameworks
Former Director, Indian Space Research Organisation, India

Preface

Federal governments globally show significant interest in the role universities play in creating national wealth by providing a knowledgeable, reflective, and enthusiastic labour force, capable of adapting to dynamically changing industry needs. This places responsibility on educators, who must modify their traditional teaching practices to ensure that graduates possess up-to-date knowledge, sufficient skills, and generic competencies. These competencies will enable them to adapt easily to the fast-changing labour market in our modern, progress-driven society.

Effective teaching and learning are central to quality research and innovation. This is influenced by various factors, including faculty expertise, teaching, and learning strategies, academic leadership, and external regulatory systems.

In response to these challenges, in 2021, the Griffith School of Engineering and Built Environment in Queensland, Australia, and the Sri Sivasubramaniya Nadar (SSN) College of Engineering Kalavakkam, Tamil Nadu, India, initiated an exciting new initiative: *monthly joint education webinars*.[1] These webinars serve as a platform for educators, researchers, and industry practitioners to disseminate their knowledge and experience, sharing best practices in learning and teaching (L&T). Over the years, leading educators and practitioners from both countries have discussed various aspects of teaching, including innovative pedagogies, classroom activities, assessment practices, and industry perspectives on graduate students. While the topics and presenters varied, most presentations focused on learning, engagement, and enhancing student preparation for industry careers. The success of these monthly webinars inspired the authors to consider creating a book that could reach a wider international audience. This book aims to identify and recommend successful teaching approaches that not only foster student engagement but also align with current and future industry expectations.

This book delves into Career-Ready Education, which encompasses preparing graduates for successful and fulfilling lifelong professional journeys. It focuses on equipping individuals with the necessary skills, knowledge, and attributes to excel

[1] Link to the joint education webinars: https://www.griffith.edu.au/griffith-sciences/school-engineering-built-environment/learning-teaching/webinars-engineering-education

in their chosen careers. Career-ready graduates embody qualities such as communication skills, integrity, enthusiasm, adaptability, critical thinking, problem-solving abilities, teamwork, and a professional work ethic. These graduates are not only prepared to enter the workforce but are also positioned to thrive, continuously learn, and grow professionally throughout their careers.

Given that career-ready graduates are a national priority, this book will be essential reading for academics, researchers, industry practitioners, community partners, and policymakers worldwide. Because the book features authors from diverse backgrounds across all chapters, its content is not aimed at any specific location. While many cases and reflections originate from Australia and India, the pedagogical strategies and teaching practices presented in this book are transferable and applicable across the majority of disciplines offered at the university level worldwide. The editors believe that this book will particularly appeal to a new generation of educators who prioritize student engagement and learning experiences. It offers practical guidelines on transforming traditional practices into more engaging methods that lead to the development of career-ready graduates.

The book is organized into 13 chapters, grouped according to four key themes:

Theme 1: A Firm Foundation: Cultivating Problem-Solving and Professional Skills for Career Success (Chapters 1–4)

Chapter 1 introduces an intervention program called the Writing Circle, designed to enhance students' written and oral communication skills through peer review and feedback. The authors share valuable insights gained from implementing the program and conclude with plans for future iterations.

In Chapter 2, the authors highlight the benefits of using creative and innovative strategies to teach English in technical academic settings. They advocate for an immersive learning environment and the application of these innovative approaches, which significantly enhance student motivation for learning English.

Chapter 3 introduces gamified and adaptive pre-laboratory modules for science courses. This innovation equips students with essential skills and competencies crucial for today's dynamic job market, fostering adaptability, problem-solving abilities, and technological fluency.

In Chapter 4, the authors present engineering classroom examples and discuss innovative strategies that help students develop the problem-solving skills required by the industry. These strategies include trial and error, logical reasoning, representation and modelling, and analysis, among others.

Theme 2: Preparing for Launch: Career Readiness in Practice (Chapters 5–8)

Chapter 5 describes the concept of immersive education, which creates a student-centred environment designed to stimulate learning. The benefits of active learning are also discussed in the context of immersive engineering laboratory exercises and project management. The chapter explores techniques to equip graduates with career-ready skills.

Chapter 6 provides readers with tools and strategies to implement innovative curricula in the engineering classroom and share lessons learned via the Scholarship of Teaching and Learning (SoTL)-focused journals and conference proceedings.

In Chapter 7, the authors showcase the portfolio assessment and reflective practice in three final-year work placement/practicum courses within the STEM disciplines, along with an associated co-curricular professional development program. The chapter demonstrates how evidence-based articulation of skills through reflection facilitates valuable skill transfer to unfamiliar contexts.

Chapter 8 explores the effectiveness, inclusivity, success, and efficiency of Work Integrated Learning (WIL) using a case study approach. It identifies trends and gaps across various disciplines and discusses feedback from key stakeholders, including educators, students, and industry partners.

Theme 3: Creating Synergies in Education and Practice: Industry Insights on Career Readiness (Chapters 9 and 10)

Chapter 9 presents and analyses feedback from industry partners regarding the readiness of final-year students to undertake industry projects. It identifies mismatches between the students' knowledge and skills, and industry expectations.

In Chapter 10, the author shares personal reflections and professional insights from the industry, along with practices for aspiring future leaders, drawing on his extensive leadership experience. Using universal case studies, the chapter discusses the importance of future-ready values and institutions.

Theme 4: University Reimagined: Embracing Change and Innovation in Higher Education (Chapters 11–13)

Chapter 11 explores the role of universities in addressing collective action problems like climate adaptation. Through a meta-analysis study, the authors discuss how sustainable universities can serve as vehicles for implementing curricula, practices, and employability strategies based on these principles.

Chapter 12 provides insights for developing future universities by exploring various scenarios that consider technological advancements, societal changes, environmental shifts, and geopolitical developments. The author presents different perspectives on the impact of technology on future society and outlines potential directions for higher education. The chapter also explores the importance of Artificial Intelligence (AI) in education, emphasizing that AI can tailor educational experiences to meet individual student needs. The chapter highlights AI's critical role in modern education, from intelligent tutoring systems and automated grading to personalized learning recommendations. AI-powered virtual assistants can provide immediate feedback, answer questions, and guide students throughout their learning journeys. The chapter concludes with key traits that future-ready graduates must possess.

Finally, Chapter 13 draws on past and present experiences to suggest several changes universities should consider focusing on producing career-ready graduates. The chapter makes a case for reimagining the goals and approaches of higher education. The authors conclude by providing critical factors for academics, university leaders, professional societies, and governments to consider, as these factors can significantly impact higher education and hence the society at large.

Looking Ahead

This book explores various aspects related to the future of universities and of teaching and learning. We believe that the foundational ideas presented will remain relevant for many years to come.

AI's impact on education seems to be evolving beyond concerns over academic integrity to become an essential component of educational frameworks. This shift has the potential not only to enhance learning outcomes but also to meet industry demands, where AI skills and adaptability are increasingly valued. By integrating AI into educational models, universities can better prepare students for a workforce likely to be shaped by AI-driven technologies and processes.

Southport, Australia
Thiruporur, India
Nathan, Australia

Ivan Gratchev
Krishnasamy T. Selvan
Hugo G. Espinosa

Acknowledgements

The editors sincerely thank the authors of this book for sharing their insights, expertise, and remarkable stories. Their contributions have been instrumental in shaping this book and bringing it to fruition.

We extend our heartfelt gratitude to the School of Engineering and Built Environment at Griffith University, Australia, and the Sri Sivasubramaniya Nadar (SSN) College of Engineering, India, for their invaluable support of the *Webinars in Engineering Education* initiative, which served as the inspiration for this book. Their encouragement and resources have been vital in enabling the realization of this project.

We express our sincere appreciation to Karthikeyan Krishnan, and Grace Liyan Ma from Springer Nature for their consistent guidance, expertise, and support in transforming the concept of this book into a tangible reality.

Finally, we are immensely grateful to our families for their support, patience, and understanding throughout the demanding process of writing and compiling this book.

In particular, Krishnasamy Selvan expresses his deep appreciation to his wife, Sudha, whose love, care, and unwavering support have been the cornerstones of his academic pursuits. He thanks his son, Shyam Krishna, often the first critic and editor of his scholarly works, for helpful discussions. He is especially appreciative of his daughter, Srinidhi, who has embraced the teaching profession and shows early signs of embodying some of the ideals discussed in this book.

Hugo G. Espinosa extends his sincere gratitude to his wife, Amanda, for her invaluable encouragement and support, which have been a driving force throughout this journey; and to his daughter, Brielle, whose curiosity and joy are a continuous source of inspiration.

Contents

Contributors

Lisa Bosman Purdue University, West Lafayette, IN, USA

Gayle Brent Griffith University, Nathan, QLD, Australia

Andrew Busch Griffith University, Nathan, QLD, Australia

Kim Cartwright Griffith University, Gold Coast, QLD, Australia

Shanton Chang The University of Melbourne, Melbourne, VIC, Australia

David Chen Griffith University, Nathan, QLD, Australia

Julie Crough Griffith University, Gold Coast, QLD, Australia

Bronwyn Disseldorp The University of Melbourne, Melbourne, VIC, Australia

Karine Dupre Griffith University, Gold Coast, QLD, Australia

Rhea Dutta Purdue University, West Lafayette, IN, USA

Hugo G. Espinosa Griffith University, Nathan, QLD, Australia

Rinske Ginsberg The University of Melbourne, Melbourne, VIC, Australia

Srikant Gollapudi School of Minerals, Metallurgical and Materials Engineering, Indian Institute of Technology Bhubaneswar, Bhubaneswar, India

Ramabadran Gopalakrishnan Retired Director Tata Sons and Leadership Consultant, Mumbai, India

Ivan Gratchev Griffith University, Gold Coast, QLD, Australia

Sarah-Jane Gregory Griffith University, Nathan, QLD, Australia; CQUniversity, Bundaberg, QLD, Australia

Barbara Hadley Griffith University, Gold Coast, QLD, Australia

Brooke E. Harris-Reeves Griffith University, Gold Coast, QLD, Australia

Ariana Henderson The University of Melbourne, Melbourne, VIC, Australia

Simon Howell Griffith University, Gold Coast, QLD, Australia

Kailani Hubbard Griffith University, Gold Coast, QLD, Australia

Deepa Kylasam Iyer School of Industrial and Labor Relations, Cornell University, Ithaca, NY, USA

Abdullah Karaksha Griffith University, Gold Coast, QLD, Australia

Shreepad Karmalkar School of Electrical and Computer Sciences, Indian Institute of Technology Bhubaneswar, Bhubaneswar, India

Martha Karunakar Sri Sivasubramaniya Nadar College of Engineering, Kalavakkam, Tamil Nadu, India

Julius Kerstien The University of Melbourne, Melbourne, VIC, Australia

Francis Kuriakose Impact and Policy Research Institute, New Delhi, India

C. Lakshmi Department of English, College of Engineering Guindy, Anna University, Chennai, India

Adrian McCallum The University of the Sunshine Coast, Sippy Downs, QLD, Australia

Antonette Mendoza The University of Melbourne, Melbourne, VIC, Australia

Brendon Munge The University of the Sunshine Coast, Sippy Downs, QLD, Australia

Andrew G. Pearson Griffith University, Gold Coast, QLD, Australia

Kimberley Reis Griffith University, Nathan, QLD, Australia

Ting Ren Griffith University, Gold Coast, QLD, Australia

Vikki Schaffer The University of the Sunshine Coast, Sippy Downs, QLD, Australia

Krishnasamy T. Selvan Department of Electronics and Communication Engineering, Sri Sivasubramaniya Nadar (SSN) College of Engineering, Kalavakkam, Tamil Nadu, India

Raghunath K. Shevgaonkar IIT Bombay, Mumbai, India

Mary-Ann Shuker Griffith University, Gold Coast, QLD, Australia

Sebastian Smith Griffith University, Gold Coast, QLD, Australia

Anne Venables The University of Melbourne, Melbourne, VIC, Australia

Jennifer C. Wilson Griffith University, Gold Coast, QLD, Australia

Chapter 1
The Writing Circle—Making Professional Skills Accessible to Students

Antonette Mendoza, Anne Venables, Shanton Chang, Ariana Henderson, Bronwyn Disseldorp, Rinske Ginsberg, and Julius Kerstien

Abstract To solve industry problems in the 21st Century, Engineering and Information Technology (IT) graduates need to be creative problem solvers, technically competent and effective communicators. In their professional lives, they must assess complex situations, devise possible options, pitch and defend their solutions, and write technical reports and software development documentation. It is incredibly challenging for educators to build the necessary communication skill sets whilst developing students' technical capabilities, particularly in coursework programs with an increasing number of students, many of whom are still developing their English language proficiency. To address this need, an intervention program called the Writing Circle 1.0 was designed and successfully trialled in 2019. The program centred around reciprocal peer review and feedback to encourage students to practise and advance their communication skills. Student feedback was positive, with students reporting benefits to their writing and improved connections with peers (Mendoza, A., Chang, S., Elliott, K. & Venables, A. in The Writing Circle: A peer-based collaborative approach to improving engineering students' scholarly writing skills. Proceedings of the 30th Annual Conference of Australasian Association for Engineering Education (AAEE 2019), Brisbane, Australia, 2019). In response to the pandemic, Writing Circle 2.0 was created as four interactive, self-paced online modules that guide students in effective writing and provide opportunities to test and apply their skills. This chapter relates the two iterations of the Writing Circle, the lessons learned in implementing each, and concludes with plans for future iterations.

1 Introduction

The demand for a skilled and technically competent workforce has become more pressing in the 21st century due to the explosion of advances in intelligent computing, artificial intelligence, robotics, and developments in biotechnology,

---

A. Mendoza (✉) · A. Venables · S. Chang · A. Henderson · B. Disseldorp · R. Ginsberg · J. Kerstien
The University of Melbourne, Melbourne, VIC, Australia
e-mail: mendozaa@unimelb.edu.au

I. Gratchev et al. (eds.), *Transformative Approaches to Career-Ready Education*,
https://doi.org/10.1007/978-981-96-7018-5_1

including advanced gene manipulations and mRNA technology, and nanotechnology, to name but a few. To respond and capitalize upon the plethora of challenges brought about by such technologies, society looks to its scientists, engineers, and Information Technology (IT) specialists seeking considered and appropriate solutions and for these experts to provide leadership and guidance when communicating their insights and possible strategic solutions (Fajaryati et al., 2020; García-Pérez et al., 2021; QILT, 2024). Therefore, these professionals must be strong and effective communicators.

It is equally essential for educators to prepare students with the necessary professional communication skill sets to ensure technical competencies (Mahmud & Wong, 2022; Souppez, 2023). In this ever-changing landscape of new and complex technological advances, graduates must communicate with various stakeholders and differing audiences (Scott et al., 2017). The difficulty for instructors across any academic program is how best to develop their students' abilities to appraise information critically, make competent decisions, and disseminate solutions through professional writing and verbal communication skills (Riemer, 2007). In programs already crammed with specialist technical knowledge, there are many questions to resolve as to how best to tackle this issue: Should instruction and communication skills development be the province of specialist language teachers? Should non-technical, complementary units be created to develop these skills, or should all instructors incorporate communication skills within their offerings, resulting in a patchwork approach where, hopefully, students will be exposed to gaining the necessary skills across their studies? Noting that in coursework programs, an increasing number of students are still developing their English language proficiency during their studies.

Increasingly, Engineering and IT academics in the coursework Masters programs at the University of Melbourne observed that their students needed further support in (1) analyzing and assessing existing published articles; (2) synthesizing and critical thinking; (3) writing up a conceptual piece of work; (4) presenting a new piece of work as part of a professional portfolio; and (5) pitching, marketing and defending opinions/solutions to problems, all of which are vital industry skills today. Moreover, this need was seen across all subjects that involved research, analytical and critical thinking aspects to solving industry problems and report writing. While the University of Melbourne provides communication skills development via a dedicated team of Academic Language and Learning (ALL) advisers, these skills are best developed within the disciplinary context (Arkoudis, 2014). Thus, a discipline-specific solution was identified to address these concerns.

In 2019, a diverse team of academics and learning specialists designed the Writing Circle initiative to target and improve graduate IT coursework students' written and oral communication skills. The intervention aimed to provide an environment and opportunities for participants to share similar experiences and learn while giving and receiving feedback from instructors and each other. The challenge in designing the Writing Circle program was how best and when to provide these opportunities for students. So, insights were sought from the education literature.

2 Background

There is a growing swell of empirical evidence, as summarized by Quaye and Harper (2014, p. 2), that university students who consistently engage in meaningful, "academic, interpersonal, and extracurricular offerings on campus" reap many rewards, including cognitive and intellectual skill development, accrual of social capital, positive images of self and improved academic performance. Optimal learning environments are those in which students feel connected and respected and can actively take responsibility for their learning (Fanghanel et al., 2016; Pittaway & Moss, 2013). This is especially true for international students who face the additional challenges of language and culture differences (Glass et al., 2015; Krause, 2005; Metro-Roland, 2018). Therefore, to capitalize on these reported positives, it was decided that students should voluntarily engage with the Writing Circle and that the program would be offered as a lunchtime activity outside of timetabled classes.

The writing research literature highlights reciprocal peer review as a powerful way of encouraging students to practice writing and to use peer feedback to advance their writing skills (Falchikov & Goldfinch, 2000; Mowl & Pain, 1995). It is argued that exposing students' work to their peers encourages them to put more effort into their writing (Jonassen, 1996; Venables & Summit, 2003). Reciprocal peer review allows a student to take on the role of a writer and reviewer. Also, reciprocal peer review is considered adequate as a type of peer-based collaboration because students working alone are unlikely to detect their own mistakes (McCarthy, 2017). By integrating student learning with peer mentoring strategies in an out-of-classroom setting, a positive and supportive environment would be established for students to obtain feedback from their peers and teaching staff (Falchikov & Goldfinch, 2000; Hardy & Bryson, 2016; Sondergaard & Mulder, 2012). This kind of self-assessment and peer review offers students a range of benefits, including modelling examples of good writing, reinforcing the importance of communication skills within the discipline, and increasing students' responsibility for developing these skills (Arkoudis, 2014). It also sends a clear message to students regarding the importance of writing practices rather than simply focusing on the product and its content (Goldsmith et al., 2019).

3 Writing Circle 1.0

3.1 Considerations

The Writing Circle initiative was designed to remediate, support, and enhance students' abilities to critically appraise information and apply professional writing skills across all subjects. The aim of the initiative was threefold: (1) to improve students' writing skills; (2) to offer an opportunity to bring along authentic assessments and receive real-time feedback from both peers and academics; and (3) to bring students together to create a sense of belonging on campus—a place to share similar experiences and learn from each other. To achieve these aims, a five-week program was structured to target graduate IT coursework students' written and oral communication skills in their first semester on campus, where they could meet and form social relationships with fellow students experiencing similar challenges. The materials and activities centred on creating opportunities and encouraging peer collaboration amongst attendees, with students engaging in reciprocal peer review and taking on the roles of writer and reviewer. Also, the program provided engagement and interactions between peers and academics outside of the regular classroom and with invited experts.

Notably, the content needed to be authentic and relevant to students. Collectively, the five modules introduced the Writing Circle and its peer-learning approach. They showcased examples of good and bad writing, deciphering rubrics and assessing others' work and their own in preparation for upcoming assessments. The five sessions were scheduled over two months during the mid-semester break. This arrangement gave participants time to settle into their new postgraduate courses and an opportunity for students to better prepare for their end-of-semester assessments.

3.2 Modules

3.2.1 Module 1—What is Good Writing?

The first session introduced students to the program goals and asked them to select an assessment task from their masters subjects that each participant wished to improve on. Students engaged in discussions on how to write a report and, through this, got to know their classmates. The program director and an expert from the University's Academic Skills team presented the session.

3.2.2 Module 2—Peer-Learning Approaches

The second session focused on student writing skills, illustrating different types of writing with examples. It began with an explanation of the fundamentals of peer-learning approaches and ended with opportunities for students to practise defending their opinions. Complementary activities gave students practice identifying examples of good writing.

3.2.3 Module 3—Rubric Writing Skills

The third session concentrated on rubrics and their use in assessing writing. In teams, students explored a rubric for the chosen assessment and learnt to assess other people's work based on the given rubric. Teams also wrote a rubric to assess a report, practising their critical thinking and writing skills in the reciprocal peer review process. Then, in teams, students presented their rubrics and provided constructive feedback on other teams' rubrics in a collaborative effort to achieve consensus on the final version.

3.2.4 Module 4—Peer-Based Teaching and Feedback

The fourth session centred on peer assessment of students' previously assessed work. As a team, students presented their rubrics and provided constructive feedback on other teams' rubrics collaboratively to achieve consensus on the final version.

3.2.5 Module 5—Report Writing Practice

The fifth session asked participants to bring one of their ongoing assessment pieces for peer review, critical analysis, and feedback. Students practised providing constructive criticism and feedback for others using peer-teaching approaches while receiving mentoring and feedback on their reports and efforts.

The inaugural Writing Circle pilot was implemented in Semester 1, 2019, supported by a $24,000 University of Melbourne Learning and Teaching initiative. The program was open and actively advertised to a cohort of over 400 students. Attendance was voluntary and varied over the different sessions (a maximum of 48), with at least 70 students participating at some stage.

3.3 Evaluation of Writing Circle 1.0

To evaluate the initiative and understand if Writing Circle 1.0 had met its aims of improving students' written communication skills whilst helping to create a sense of community, feedback from the participants was sought after each session querying:

- perceptions of the program's value in helping develop their technical writing skills
- an assessment of the degree to which peer collaboration helped to build a learning community for participants
- identification of the weekly activities most useful in helping students understand and analyse arguments, critical thinking and report writing
- overall satisfaction, or otherwise, with the Writing Circle program
- suggestions for improvement to the program

Attendees were asked to complete an anonymous survey of 19 questions, of which 16 were statements relating to the session's activities. Following these statements, two survey questions asked for a written response, and a final question asked if the respondent's participation in the session was either more, the same, or less than in previous sessions. Ethics approval was sought and granted to collect these anonymous survey responses from students.

The collation and analysis of student responses were done first on a weekly basis and then across the initiative. A comparison showed progressively more students self-reporting that their participation was increasing; perhaps this was not surprising given that the initiative required more student engagement with their own assignments in the later sessions.

Regarding content and presentation, students found the Writing Circle 1.0 program:

- design helps them learn and understand the topic
- materials were presented clearly by knowledgeable speakers
- (increasingly more) engaging, thoughtful and provoking

Further, there is strong support that the cohort believed that the sessions:

- helped in organizing their thoughts and structuring their writing by reading others' work
- enabled connections with others in the class
- were relevant and worthwhile to current and future studies and projects

The feedback helped identify activities, like peer review and understanding assessment rubrics, that students found helpful in developing understanding and skills in analysis of arguments, critical thinking and report writing. Also, participants were more appreciative of a reviewer's perspective and then more willing to accept critical feedback. Open-ended question responses affirmed the positive value of and students' support for engaging with peers and academics outside the classroom. Further details of the pilot, the survey instrument, and feedback collection and collation are reported in Mendoza et al. (2019). This paper adds to the limited yet growing body of case studies in collaborative peer review in computing and engineering (Carlson & Berry, 2008; Sondergaard & Mulder, 2012).

Participants' feedback endorsed the Writing Circle approach, leading to plans to continue the program. With the next iteration scheduled for Semester 2, 2020, we planned a revision of the materials and activities to address three aspects of Writing Circle (1.0) seen as requiring further improvement and enhancement:

- Skills development
    - seek methods to measure students' actual performance in critical appraisal and professional writing skills
    - further improve student oral communication and technical presentation skills with specific training and practice to build confidence
- Delivery mode
    - research possible strategies to optimize delivery approaches
    - identify the most appropriate staff for each session
    - investigate the timing of the program modules in relation to student workloads, avoiding major assessment tasks

- Participation rates
  - increase the number of students engaging with the program

4 Writing Circle 2.0

The period of our evaluation and revisions of Writing Circle 1.0 in 2020 coincided with the pandemic. In response, the University offered students dual-mode classes: a mix of campus-based face-to-face teaching (f2f) with synchronous online participation by remote students participating in blended learning environments where online learning materials and activities were part of the subjects' requirements.

The disruptions caused by the pandemic to usual teaching modes diminished opportunities for social engagement among students and their teachers. Like their contemporaries worldwide (United Nations, 2020, 2021), our students felt isolated and disconnected from their peers, subject content, and staff. As educators unable to schedule face-to-face sessions with our students, we needed to re-imagine how best to offer Writing Circle 2.0 to students. We committed to improving written and verbal communication skills development, increasing the focus on presentations and pitching ideas, and increasing student participation opportunities. We envisaged a set of interactive and reusable modules incorporating the most effective components of Writing Circle (1.0). This would enable a shift from solely face-to-face delivery to online during the pandemic and move post-pandemic towards a blended engagement model incorporating these online components. This shift towards online modules would enable students to engage in the program at critical points in their first semester at their own pace and convenience.

4.1 The Approach

The approach was to craft four interactive, self-paced online modules encompassing activities to present students with a range of writing tasks, guidance for effective writing, instruction and practice on oral presentations, and opportunities to test and apply their skill development.

1. **Writing effectively**: critical thinking practice introducing various techniques for improving the clarity of writing.
2. **Peer collaboration and feedback**: techniques to offer effective feedback to others on the aspects of writing that make the most significant difference.
3. **Understanding rubrics**: illustrates different ways student writing may be assessed, with examples of various marking criteria presented so students can reflect on how to improve their submissions and provide helpful peer assessments.
4. **Presenting Effectively**: Illustrates the Importance of Visual and Non-Verbal Cues in Oral Communications.

We proposed that these interactive modules incorporate role play, scenario-based, and video-enhanced case studies to engage students with each other while practising their written and presentation communication skills. For each module, an appropriate syllabus was discussed and decided with engineering and IT academics, a specialist senior learning designer from the University's Learning Environments team, a teaching consultant from the Academic Skills team, and performance and presentation experts from the Victorian College of the Arts. Details of each module follow. Some reorganization and alignment of the modules presented in Writing Circle 1.0 was needed due to changes of emphasis between programs and sometimes due to the constraints of Writing Circle 2.0 being online.

4.1.1 Module 1: Writing Effectively

In developing the module, the focus was on techniques that students could a) easily follow without further instruction, b) apply to any genre of writing, e.g. reports, case studies, emails, reflective journals, and c) carry with them into their professional work after graduation. The goal was to equip students with the ability to:

- Recognize critical features of clear sentences and paragraphs, and
- Identify and select specific techniques to enhance clarity in writing.

The approach closely replicated that of the Writing Circle 1.0 in-person workshop by teaching students the key features of clear writing before selecting specific techniques to enhance their writing clarity. Throughout the module, students are invited to apply the techniques to sample writing and their writing. Commencing the online module, students are asked to write a sample paragraph and put it aside for later before being introduced to the following five writing strategies to improve the clarity of a sample paragraph immediately.

1. Introduce ideas
2. Move from general information to more specific information
3. Put old or given information before new information
4. Include a link to the previous sentence
5. Keep the subject and the verb at the beginning of the sentence

To practise applying each of the above techniques, students progressively work through two instances of technical writing, answering quiz questions as they progress (Fig. 1). These opportunities to practice help students come to their understanding of how easy the strategies are to implement and the impact it has on their final product. The process also helps students internally question their assumptions about the objectivity (and perceived subjectivity) of clear writing and increase self-awareness about their writing practices. Finally, students are guided step by step to apply the five strategies to improve their original paragraph and submit an online checklist to identify opportunities for improvement (Fig. 2).

The module scaffolds the students' learning by stepping through the techniques slowly. It helps build confidence in their ability to see the larger picture and feel

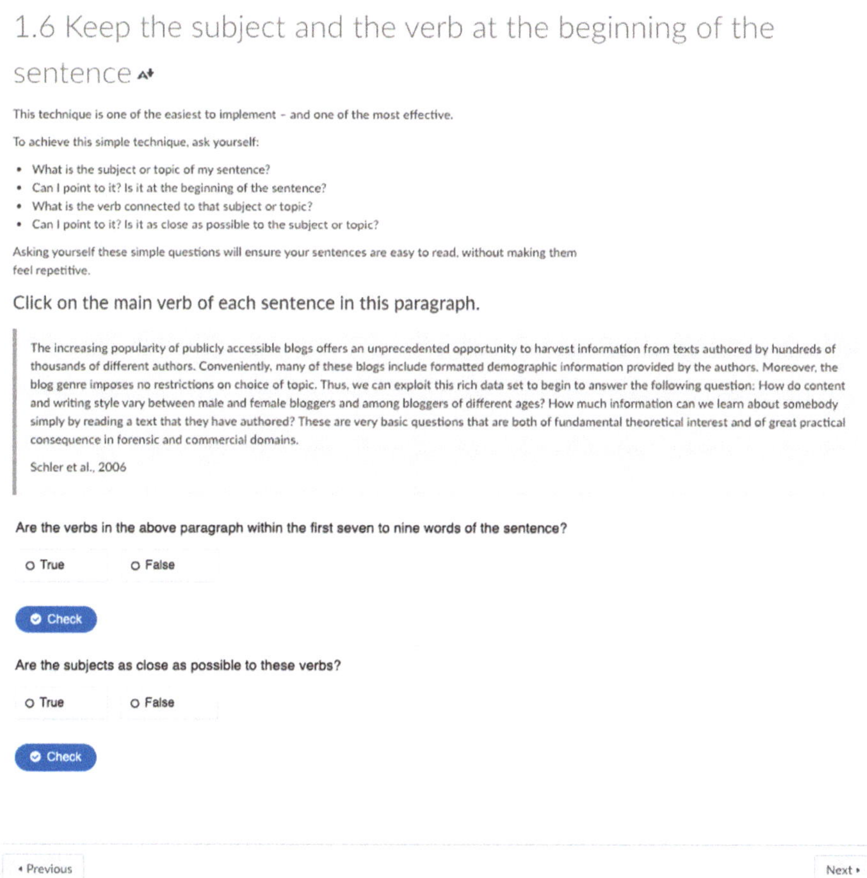

Fig. 1 Sample technical writing exercise with quiz questions

capable of independently applying the techniques. The sense of community and belonging achieved in the in-person Writing Circle 1.0 is replicated in this online module, as students see their peers' attempts at an improved example only after submitting their own attempts.

The strategies introduced in the module can apply to any genre of writing, thereby encouraging students not to get caught up in the type of writing but instead focus on how clear it is. The value of clear writing is normalized, as all students can benefit from improving the clarity of their writing, regardless of their level of language proficiency.

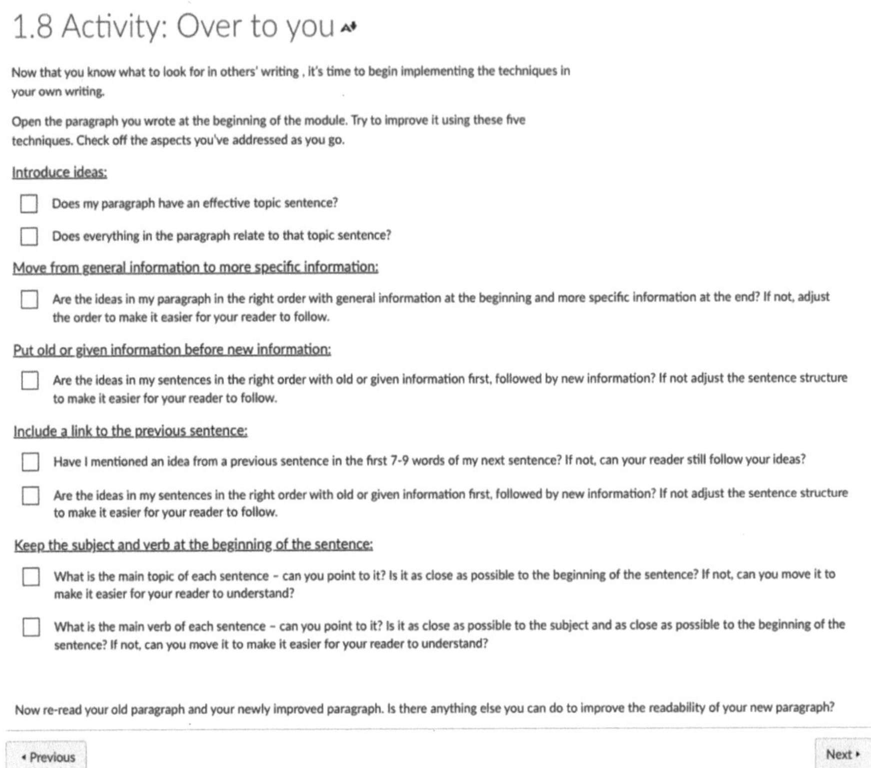

Fig. 2 Sample activity guiding students in applying the five strategies to improve their original paragraphs

4.1.2 Module 2: Peer Feedback

The focus of this module was peer feedback: its importance, benefits, and approaches. The goal was to assist students in:

- Identifying elements of effective feedback
- Differentiating between editing and proofreading
- Offering considered feedback on the aspects of writing that make the most difference

Similarly, this module replicates the approach taken in the peer-learning session of Writing Circle 1.0 by offering opportunities to practice feedback techniques alongside interactive activities and discussion boards to help build a sense of community and belonging. Firstly, students are invited to share a time when they received helpful feedback and identify some possible benefits of giving feedback to others. By submitting their own experiences, students can then read the insights shared by others, building a bank of diverse perspectives to learn from. A short video was developed to teach students the following five key tips for giving better feedback:

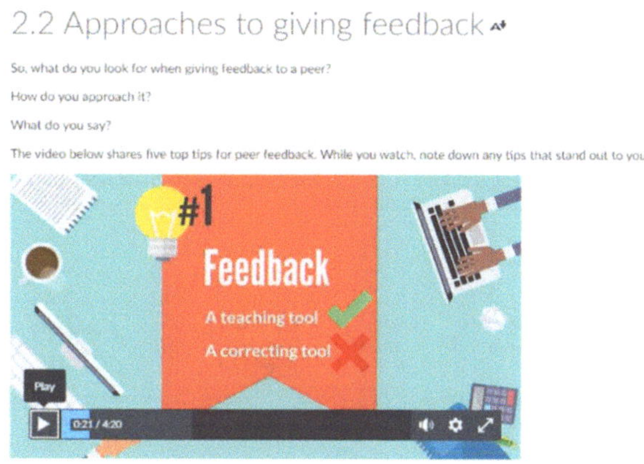

Fig. 3 Video to gain knowledge on how to provide peer feedback techniques

1. Focus on teaching rather than correcting.
2. Read carefully to identify high-impact improvements.
3. Focus on identifying improvements in structure and linking
4. Focus on the big picture rather than the minor issues
5. How you give feedback is just as important as what you are feeding back on.

Of these, students are encouraged to take note of the strategies that most resonate with them in much the same way as they would during the in-person session. Then, through online activities, students test their understanding of peer feedback techniques (Fig. 3). For example, students attempt to separate the common issues in writing into editing and proofreading categories to confirm their understanding of minor issues compared with those that will significantly affect the overall paper. Finally, students attempt to apply this approach to sample paragraphs as they would have done during the in-person session.

While progressing through the module, students are encouraged to focus on the issues that make the most impact to deliver effective feedback manageable for the writer (Fig. 4). In designing the module, the advantages of working with others when reviewing and improving writing were a key focus. Like the in-person workshop of Writing Circle 1.0, this online module aims to leave students with a sense that their writing can also improve by engaging in peer review and feedback on others' writing. As students apply the techniques learnt during the module, they are more likely to recognize issues in their own writing both at university and in the future.

4.1.3 Module 3: Rubrics

The module builds on the effective writing skills learnt in Module 1 and peer feedback practices introduced and practised in Module 2.

This module prepares students to a) understand how rubrics communicate task requirements and b) use rubrics to help develop an understanding and analysis of

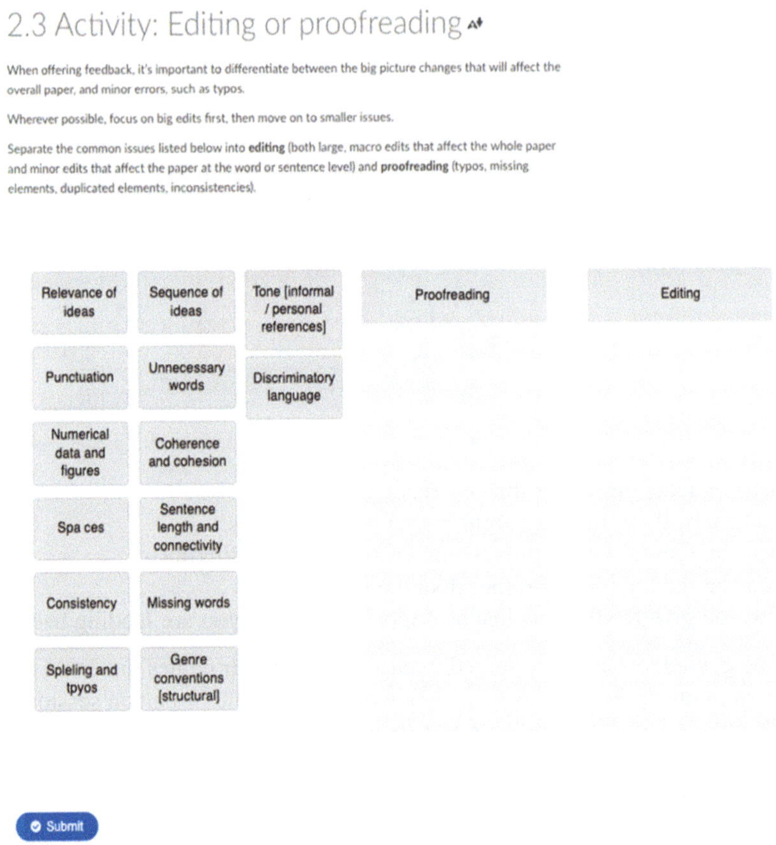

Fig. 4 Activity on editing and proofreading

arguments, critical thinking, and report writing. The goal of the module was to assist students to.

- Identify different kinds of rubrics
- View and use feedback provided through a rubric
- Understand how markers use criteria descriptions and weighting to calculate a final grade.

Mimicking the approach in Writing Circle 1.0, this module illustrates how markers may assess student writing and the feedback provided using a rubric structure. Since students have varying amounts of prior experiences of using rubrics, examples of different types of schemes and marking criteria are presented to build familiarity with the rubric tools and invoke students' reflections (Fig. 5). The module includes a mixture of explanatory text annotated illustrations, links to relevant user guides, and reflective activities to nurture participants and improve their future written submissions and understandings of critical feedback.

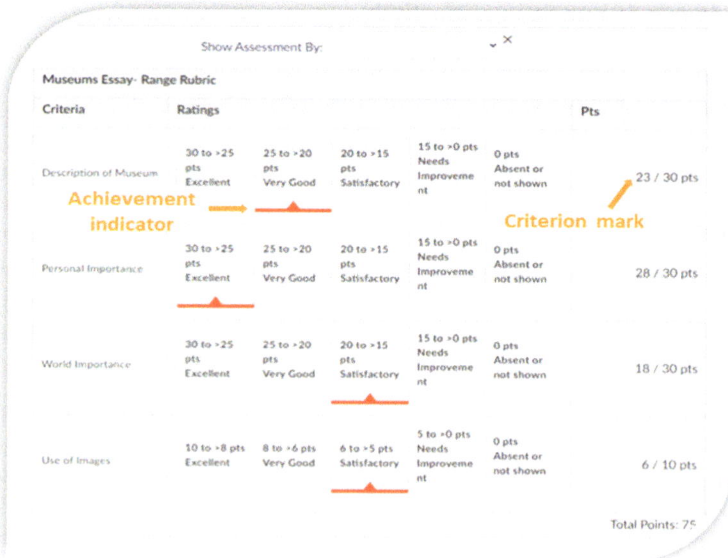

Show Assessment By:

Museums Essay- Range Rubric

Criteria	Ratings					Pts
Description of Museum	30 to >25 pts Excellent	25 to >20 pts Very Good	20 to >15 pts Satisfactory	15 to >0 pts Needs Improvement	0 pts Absent or not shown	23 / 30 pts
Personal Importance	30 to >25 pts Excellent	25 to >20 pts Very Good	20 to >15 pts Satisfactory	15 to >0 pts Needs Improvement	0 pts Absent or not shown	28 / 30 pts
World Importance	30 to >25 pts Excellent	25 to >20 pts Very Good	20 to >15 pts Satisfactory	15 to >0 pts Needs Improvement	0 pts Absent or not shown	18 / 30 pts
Use of Images	10 to >8 pts Excellent	8 to >6 pts Very Good	6 to >5 pts Satisfactory	5 to >0 pts Needs Improvement	0 pts Absent or not shown	6 / 10 pts

Total Points: 75

Fig. 5 Example rubric viewable after grading

To present the various aspects of rubrics and their role in guiding student writing and shaping assessment and feedback, the following steps are taken throughout the module:

1. Introduce an outline of the different forms of rubrics and possible roles in assessment and feedback
2. Student reflections on previous experiences with rubrics and marking schemes through responding to questions in a voluntary and ungraded survey
3. Viewing rubrics common in the University's learning management system (LMS)
4. Rubrics in other learning technologies, e.g. Feedback Fruits
5. Tips for using rubrics in assignment preparation and after receiving feedback
6. Practice grading writing samples to a given rubric.

Importantly, student exposure and familiarity with the University's LMS and other online technologies, such as Feedback Fruits, is promoted as it is likely that during their courses, students will be required to complete similar tasks using structurally different tools. In this module, students use peer review when they assess and give feedback on submissions by others. Participants reflect and provide feedback on the contributions and behaviours of other students working together on a group assignment.

4.1.4 Module 4: Presenting Effectively

In our examination and review of the pilot Writing Circle 1.0, we identified a strong emphasis on good written communication skills and noticed the omission of oral communication practice. To redress the imbalance, we consulted with the Victorian College of the Arts (VCA) performance specialists on appropriate strategies and approaches for improving student oral presentation skills. As the module's goal was to support students in becoming more effective communicators and presenters, these specialists proposed that an understanding of the important elements of communicating effectively through self-awareness and non-verbal communication could help improve students' oral communication. Suggested strategies to achieve this would include:

- Identification and awareness of elements of non-verbal communication in everyday life,
- Play and practice of non-verbal elements to optimize presentations in a range of interpersonal, academic and professional contexts,
- Use of breathing techniques to calm nerves, centre oneself and improve communication, and
- Education and adoption of behaviours in specific teaching and learning contexts must be practical and engaging.

As there was no equivalent of this module in the pilot program, we had intended to have VCA specialists run interactive webinars with students using performance strategies, such as body language, voice training and space utilization during presentations as part of the Writing Circle (2.0) iteration. However, restrictions due to the pandemic and our move to online modules meant small video clips of excellent and poor efforts were filmed and added as part of the instruction throughout the module. It was designed to be a comprehensive resource for students to consider how our whole bodies communicate and offer suggestions for improving their practice and becoming more effective communicators.

The methods and exercises used for the online version included:

1. Introduction to effective communication, particularly non-verbal aspects, for consistency between what is said in words with non-verbal communication given by a speaker's body,
2. Information and activities to explore body language concentrating on gaze, gestures, posture
3. Examination of status and the impact of posture, and the use of space: position, and proximity,
4. Discussion and exploration of energy types: present, introverted, and extroverted,
5. Control of breathing, diaphragmatic breathing practice,
6. Discussion and practical guide on preparing for making presentations and
7. Summary and reflection exercises

As students engage with the above steps by working through the information, videos, activities, and quizzes, their observational, analytical, physical, and vocal skills are expected to be bolstered and improved. By observing others, students can identify and adjust aspects of their body language appropriately to support their presentations.

4.2 *Implementation of Writing Circle 2.0*

Once the four modules were developed, a discussion ensued regarding the most appropriate platform to host Writing Circle 2.0 in its full online mode. In this iteration, we sought to make Writing Circle 2.0 accessible to students from any discipline across the University, whether remaining on campus or studying online. The University's student learning management system, Canvas, was selected to offer all the technical components necessary to securely allow participant interactions whilst enabling compliance with various University online security and privacy policies. By making the Writing Circle a Canvas community, the modules are available to all university students and staff; Canvas users can enrol on the community page, which includes the four modules for free on the University's subjects and communities' catalogue through a dedicated link (Fig. 6). Additionally, students can conveniently access the modules within the same platform where they engage with their subjects allowing participants to review the Writing Circle content throughout their degree.

Subsequently, each of the four modules was developed using H5P plugins. The suite of Writing Circle modules was uploaded to Canvas early in March 2021, where it has remained available to the University community.

To indicate participants' engagement and activities with each module, routine internal analytics within Canvas, such as Log files, can be paired with analytics from the H5P activities embedded in modules. In addition, upon module completion, users were asked to complete a voluntary and anonymous Qualtrics survey about their experiences with the materials. Approval was obtained from the University for Human Research Ethics for these surveys. Participant consent was collected via a preface page to the survey. All data collected has been stored in compliance with the University's Management of Research Data and Records Policy (MPF1242).

4.3 *Evaluation of Writing Circle 2.0*

The first step in evaluating Writing Circle 2.0 involved an examination of various log file records. These records indicate that more than 3000 individuals accessed one or more pages of the four modules over two and a half years. An inspection of access per individual page indicates that most participants visited various pages more than once. The mode for page access per participant is five visits, and the median is eight engagements per participant. As expected, the introductory home page of Writing Circle 2.0 was accessed most often, with nearly 4000 viewings.

Fig. 6 Landing page for the writing circle modules

The components supporting student interactions within modules were created using various H5P plugins, and the associated H5P reports can help measure traffic and communications by participants on a per-module basis. The log files attest that many students visited the hosting pages for each activity. However, paradoxically, the H5P reports show that many students skipped the interactive activities and travelled through the hosting page and module. For instance, in Module 1: Writing effectively, 53 students made up to 3 attempts to work towards a correct solution in activity 1. This reduced to 33 students, making up to two attempts at a correct solution for activity 2. This was reduced further in Module 2: Peer Feedback, where the H5P reports show fewer logins (14 students) engaging up to three times per login. This shows that each participant persevered until they achieved their maximum possible score. By Module 4: Presenting effectively, response numbers dwindled further, with only two students recording a response to the final activity.

The two activities of Module 3 showed poor participation numbers, with nine students providing written feedback in the first activity and only three responding in the second and third. From those who engaged with the first activity when asked about their previous rubric experiences, several considered answers were given, such as, *"(I) read it before submitting assignments to check that I have covered the points outlined in the rubric."* and *"most rubrics I have seen use several columns to give an example of the indicative feedback for each mark band"*. Likewise, those who answered the second activity noting their use of rubrics mentioned, *"I only used it at the very end right before submitting my work, making sure that my work satisfied the necessary criteria"* and *"I always read available rubrics and marking schemes before starting the assignment."* Importantly, the third activity asked students to assess an example with a given rubric. The assessments by participants are encouraging to the online writer, suggesting they *"add topic sentence and conclusion may be easier to follow, more reference and with comparing more data handling method"* and *"Your introduction does not really address the topic that you examine later in the essay. You do not explain what kinds of applications may prefer cloud or edge"*. These assessments illustrate students' incorporation into their suggestions of skills learnt in Module 1: Writing Effectively and Module 2: Peer Feedback.

The next step in the evaluation process of Writing Circle 2.0 was to inspect the anonymous survey feedback for each module. Students needed to reach the final page of the module to be invited and give informed consent to a module's corresponding survey. The Qualtrics surveys comprised several statements for agreement or otherwise, with freeform questions about what students found interesting, how the module could be improved, and any other comments or questions participants would like to make. Although slightly different, the surveys were an abbreviated version of the 19-question survey used to evaluate Writing Circle 1.0 published by Mendoza et al. (2019).

Due to limited survey responses, the data was insufficient to identify specific trends or compare Writing Circle 1.0 and Writing Circle 2.0. Despite the low response rate, most of the students who did respond indicated they agreed or felt neutral about the survey statements. Furthermore, students shared the following feedback in response to the Writing Effectively and Rubrics module surveys:

Writing effectively

- *"Writing a paragraph and going back to review it helped improve my writing."*
- *"Explaining the difference between proofreading and editing was helpful" and "More practice questions would be helpful."*
- *"Sentences organized from general information to more specific information is much easier for readers to follow." and "More examples and practices provided will be better."*

Rubrics

- *"The marking a sample essay activity was very helpful in understanding how marking rubrics are applied."*

There were no freeform written answers in the Survey responses for the Peer feedback or for Presenting effective surveys.

5 Lessons Learnt for the Future

Overall, the evaluation attempts of the Writing Circle 2.0 online suite have borne little fruit. This is despite the considerable data in the comprehensive log files across all modules in the suite or the interactions recorded through analytics of the H5P plugins. What needs to be improved is direct input from our students in their interactions with module activities, leading to few survey responses, particularly those giving written opinions. More data is needed to indicate if participants perceive value in the program, notice any improvements in their written or oral communication skills, or see value in peer interactions and feedback. It is impossible to hear students' voices on their overall satisfaction or otherwise if they choose to remain silent.

The log file traffic to the suite of online modules suggests that students are engaging with each of the four Writing Circle 2.0 modules. However, the lack of feedback from other participants could be due to a few factors. Possibilities include:

- Students are more motivated to complete tasks directly related to the assessment
- Students are more motivated to complete tasks with others rather than in an automated online environment
- Students choose to prioritize their studies over completing surveys with no direct feedback.

Feedback from the successful delivery of Writing Circle 1.0 as a voluntary lunchtime activity showed that the intervention had met its aims of improving students' written communication skills whilst offering opportunities for students to receive real-time feedback in an environment that helped to foster a sense of community through learner connections with their peers and academics (Mendoza et al., 2019). The literature attests that when students feel included and connected to a group, class, subject and institution, they are more engaged in their studies and achieve better academic outcomes (Mendoza & Venables, 2023; Metzger & Taggart, 2020; Peacock & Cowan, 2019; Wilson et al., 2018; Yorke, 2016).

With the migration to online modules and making the Writing Circle 2.0 suite accessible to broader cohorts of students across multiple disciplines at the University, we assumed that students would be online synchronously and connect to peers when on the platform. This did not happen. We believe it could be (1) students feeling lost and alone due to COVID; (2) a lack of feeling connected with their peers on the platform; (3) a sense of community missing with the online-only module presentations, or (4) students being economical with their study time and acting transactionally in their studies.

In the future, we ask how to improve student participation and engagement with each module. We also seek improved strategies to collect better and more informative feedback than was the case for Writing Circle 2.0. Several possibilities to bolster student engagement could involve:

- Aligning the Writing Circle suite more closely with specific units within the teaching program so that subjects reliant on writing and presenting skills refer to and actively encourage students to visit the modules supporting communications or,
- Setting some or all module completions as prerequisites or hurdles for subjects that rely on writing or presenting skills. We wish to investigate making a hurdle requirement for students in their first semester to actively engage with a minimum number of Writing Circle (2.0) modules or,
- Tying one or more module completions to assessments in core units, we note that online Module 3: Rubrics had high logins and that assessment is a big motivator for student engagement.

In conclusion, from the lessons learnt and the positive feedback from students, we firmly believe the intervention program called the Writing Circle in the hybrid form is a way forward to enhance students' communication and presentation skills through an activity-based approach.

References

Arkoudis, S. (2014). *Integrating English language communication skills into disciplinary curricula: Options and strategies*. Department of Education, Skills and Employment. Retrieved from https://melbourne-cshe.unimelb.edu.au/__data/assets/pdf_file/0010/1805185/Arkoudis_S_NST_report_2014.pdf

Carlson, P., & Berry, F. C. (2008). Using computer-mediated peer review in an engineering design course. *IEEE Transactions on Professional Communication, 51*(3), 264–279.

Fajaryati, N., Budiyono, Akhyar, M., & Wiranto, W. (2020). The employability skills needed to face the demands of work in the future: Systematic literature reviews. *Open Engineering, 10*(1), 595–603. https://doi.org/10.1515/eng-2020-0072

Falchikov, N., & Goldfinch, J. (2000). Student peer assessment in higher education: A meta-analysis comparing peer and teacher marks. *Review of Educational Research, 70*(3), 287–322.

Fanghanel, J., McGowan, S., Parker, P., McConnell, C., Potter, J., Locke, W., & Healey, M. (2016). *Literature review. Defining and supporting the Scholarship of Teaching and Learning (SoTL): A sector-wide study* [report for University of Brighton].

García-Pérez, L., García-Garnica, M., & Olmedo-Moreno, E. M. (2021). Skills for a working future: How to bring about professional success from the educational setting. *Education Sciences, 11*(1), 27. https://doi.org/10.3390/educsci11010027

Glass, C. R., Wongtrirat, R., & Buus, S. (2015). *International student engagement: Strategies for creating inclusive, connected, and purposeful campus environments*. Stylus Publishing, LLC.

Goldsmith, R., Willey, K., & Boud, D. (2019). Investigating invisible writing practices in the engineering curriculum using practice architectures. *European Journal of Engineering Education, 44*(1–2), 71–84.

Hardy, C., & Bryson, C. (2016). The salience of social relationships and networks in enabling student engagement and success. *Student Engagement in Higher Education Journal, 1(1).* Retrieved March 13, 2018, from https://journals.studentengagement.org.uk/index.php/raise/article/view/376/335

Jonassen, D. H. (Ed.). (1996). *Handbook of research for educational communications and technology.* Simon & Schuster Macmillan.

Krause, K. (2005). Understanding and promoting student engagement in university learning communities. *A keynote address 'Engaged, inert or otherwise occupied? Deconstructing the 21st century undergraduate student' at the James Cook University Symposium 2005* (pp. 21–22).

Mahmud, M. M., & Wong, S. F. (2022). Stakeholder's perspectives of the twenty-first-century skills. In *Frontiers in Education. 7.*

McCarthy, J. (2017). Enhancing feedback in higher education: Students' attitudes towards online and in-class formative assessment feedback models. *Active Learning in Higher Education, 18*(2), 127–141.

Mendoza, A., Chang, S., Elliott, K., & Venables, A. (2019). The writing circle: A peer-based collaborative approach to improving engineering students' scholarly writing skills. *Proceedings of the 30th Annual Conference of Australasian Association for Engineering Education (AAEE 2019)*, Brisbane, Australia.

Mendoza, A., & Venables, A. (2023). Attributes of blended learning environments designed to foster a sense of belonging for higher education students. *Journal of Information Technology Education, 22*, 129–156. https://doi.org/10.28945/5082

Metro-Roland, M. (2018). Community, identity, and international student engagement. *Journal of International Students, 8*(3), 1408–1421. https://doi.org/10.32674/jis.v8i3.63

Metzger, M., & Taggart, J. (2020). A longitudinal mixed methods study describing 4th-year baccalaureate nursing students' perceptions of inclusive pedagogical strategies. *Journal of Professional Nursing, 36*(4), 229–235. Epub 2019 Dec 23. PMID: 32819549. https://doi.org/10.1016/j.profnurs.2019.12.006

Mowl, G., & Pain, R. (1995). Using self and peer assessment to improve students' essay writing: A case study from geography. *Innovations in Education and Teaching International, 32*(4), 324–335.

Peacock, S., & Cowan, J. (2019). Promoting a sense of belonging in online learning communities of inquiry at accredited courses. *Online Learning, 23*(2), 67–81. https://doi.org/10.24059/olj.v23i2.1488

Pittaway, S., & Moss, T. (2013). Student engagement in and through orientation. In E. Dunne & D. Owen (Eds.), *The student engagement handbook: Practice in higher education* (pp. 275–290). Emerald Group Publishing.

QILT. (2024). 2022 Employer satisfaction survey report: Employer views of recent graduates. Retrieved from https://www.qilt.edu.au/surveys/employer-satisfaction-survey-(ess)#anchor-3

Quaye, S. J., & Harper, S. R. (Eds.). (2014). *Student engagement in higher education: Theoretical perspectives and practical approaches for diverse ePopulations* (2nd ed.). Routledge. https://doi.org/10.4324/9780203810163

Riemer, M. (2007). Communication skills for the 21st century engineer. *Global Journal of Engineering Education, 11*(1), 89–100.

Scott, F. J., Connell, P., Thomson, L. A., & Willison, D. (2017). Empowering students by enhancing their employability skills. *Journal of Further and Higher Education, 43*(5), 692–707. https://doi.org/10.1080/0309877X.2017.1394989

Sondergaard, H., & Mulder, R. A. (2012). Collaborative learning through formative peer review: Pedagogy, programs and potential. *Computer Science Education., 22*(4), 343–367.

Souppez J-BR. (2023). Engineering employability skills: Students, academics, and industry professionals perception. *International Journal of Mechanical Engineering Education, 53*(1), 125–142. https://doi.org/10.1177/03064190231214178

United Nations. (2020, August). Education during COVID-19 and beyond. *Policy Briefing*. https://www.un.org/development/desa/dspd/wp-content/uploads/sites/22/2020/08/sg_policy_brief_covid-19_and_education_august_2020.pdf

United Nations. (2021). During the pandemic, students rely on solidarity. *Stories from the field*. Retrieved from https://www.un.org/es/coronavirus/during-pandemic-students-rely-solidarity.

Venables, A. & Summit, R. (2003). Enhancing scientific essay writing using peer assessment. *Innovations in Education and Teaching International, 40*(3), 281–290.

Wilson, R., Murray, G., & Clarke, B. (2018). The RMIT belonging strategy: Fostering student engagement in higher education. In D. Wache & D. Houston (Eds.), *Research and development in higher education: (Re) valuing higher education* (Vol. 41, pp. 257–266). Adelaide, Australia.

Yorke, M. (2016). The development and initial use of a survey of student 'belongingness', engagement and self-confidence in UK higher education. *Assessment & Evaluation in Higher Education, 41*(1), 154–166.

Chapter 2
English and Engineering Students: A Love-Hate Relationship

Martha Karunakar

Abstract The relationship between the teaching and learning of the English language and engineering students is one of love-hate relationships. On the one hand, English is indispensable for technical communication, career prospects and upward social mobility. Conversely, the students express apathy towards learning and practising language skills, particularly in technical academic environments. English is the accepted global language of communication in most professional fields. Thus, English language teaching and learning is essential due to its significance in career readiness and advancement. Challenges teachers face include heterogeneous classrooms, mother tongue interference, treating English as a foreign language, large classes, indifferent attitudes of the learners, and the tendency to learn English as another subject rather than a language. Although educators have previously emphasized learning the language by memorizing grammar rules and syntactic structures and, later, using language in real-life situations, the focus has moved to create exciting and motivating situations for the students to learn and use the language. This chapter underscores the importance of trying various materials and activities to break the monotony of conventional teaching methodology and as a supplement to the primary teaching activity.

1 Introduction

Globally, English is the most spoken language, and in India, with several students travelling abroad for higher studies and better work opportunities, proficiency in English is becoming increasingly crucial. Recent decades have witnessed extensive international student migration, and according to the 'Indian Students Mobility Report 2023' (Chhapia, 2023), the top destination for Indian students is the U.S., followed by Canada, the U.K. and Australia. Although students may meet the minimum entry requirement for English language proficiency, they may encounter challenges with the academic medium of instruction and social integration.

M. Karunakar (✉)
Sri Sivasubramaniya Nadar College of Engineering, Kalavakkam, India
e-mail: marthak@ssn.edu.in

I. Gratchev et al. (eds.), *Transformative Approaches to Career-Ready Education*,
https://doi.org/10.1007/978-981-96-7018-5_2

English is no longer solely a language for ethnic or national identification as languages are conventionally considered…it is an economic skill, a marketable commodity and a form of cultural capital" (Salomone, 2022, p. 8). David Crystal (2003) suggests that the ratio of native speakers of English to the non-native speakers is 1:3 and "the population growth in areas where English is a second language is about 2.5 times that in areas where it is a first language" (p. 69).

Proficiency in English facilitates effective communication and includes soft-skills development, critical thinking, problem-solving and teamwork, which are crucial in the globalized workspace of engineering and technology. To empower students to be successful and thrive in diverse professional environments, teachers of English underscore the practical application of English language skills within engineering contexts, thereby bridging the gap between academia and industry expectations.

To meet the shifting paradigms in terms of the expectancy of language proficiency, this chapter aims to demonstrate the advantages of applying creative and innovative strategies, materials, methods, and activities in teaching the English Language in technical academic environments. It also argues in favour of an immersive learning environment and application of innovative teaching approaches leading to increased student motivation for English language learning. The traditional English language teaching methodology focuses on developing the four skills: Listening, Speaking, Reading and Writing (LSRW), vocabulary, and grammar. To accommodate the shift in expectations regarding language proficiency, it is ideal to begin the process by framing a syllabus "structured on language functions and skills rather than language items and technical content" (Rus, 2020, p. 339).

2 National Education Policy (NEP, 2020)

India is revamping its education policy, and the Central and State governments are heavily investing in it. This pedagogic revolution regarding the National Education Policy (NEP) of 2020 is a comprehensive framework introduced by the Government of India that aims to reform the country's education system. Dr K. Kasturirangan, a space scientist and the mind behind NEP 2020, headed the drafting committee of NEP and believes that higher education must be interdisciplinary and holistic and provide access, equity and inclusion to all sections of society.

NEP 2020 promotes institutional autonomy, quality research, continuous professional development of teachers, technology integration, higher education internationalization, governance and regulatory architecture restructuring, multidisciplinary curricula, engaging blended pedagogy, etc. NEP 2020 also encourages freedom for faculty to design their "own curricular and pedagogical approaches within the approved framework, including textbook and reading material selections, assignments, and assessments. Empowering the faculty to conduct innovative teaching, research, and service as they see best will be a key motivator and enabler for them to do truly outstanding, creative work" (p. 40). It is against this backdrop that Engineering undergraduates are involved in English language teaching–learning.

3 Background of the Tertiary Level Students in India

Teaching English to the tertiary level students in the Indian context is unique, mainly because India is a multilingual country, and students come to college or university with varying abilities in multiple languages. For example, in the context where this author teaches, most students can speak 3–4 languages like Tamil, Telugu, Malayalam, and Kannada, that is, if they are from the neighbouring southern states. Students from the northern states are fluent in Hindi and other languages like Odia, Bengali, Kashmiri, etc. Although many of them can speak English fluently and coherently, some find speaking in English challenging.

Typically, there are approximately around 60 or more students in a class. There is a designated syllabus for each semester and a minimum number of 45 hours in which it is expected to be completed. English is taught as a second language. The objective of teaching English at the tertiary level for these Engineering students is not only to perform well in campus placement interviews but also to equip them with a comprehensive skill set that renders them job-ready upon graduation.

Over the years, in the context of teaching English as a second language, several language teaching models were introduced and adopted by prominent linguists and scholars like the grammar-translation approach, direct approach, cognitive approach, communicative approach, etc., some of which are retained and practised while others overlooked or neglected for various reasons. Teachers, too, have tried several creative and innovative methods of teaching English and found them to be effective. Saienko and Shevchenko (2020) introduced authentic videos in the English class and found improvement in the 'technical students' listening and speaking skills by almost two times' and the 'average total mark, comprising the outcomes of listening, reading, writing, and speaking tasks, … by 1.5 times' (p. 350). Kovalyova et al. (2016) highlight the efficacy of project-based teaching of English, with a particular focus on oral and written communication skills, to engineering students in Russian Technical universities. Another innovative approach was the Personalized English Language Enhancement (PELE) approach (Kim, 2023), which was started as an extra-curricular programme for international students and was found to enhance English speaking skills in 'both academic and everyday contexts and self-efficacy skills for self-regulated English learning and academic and social engagement'. TBLT (task-based language teaching) is an instructional approach popularised by N.S Prabhu with his Bangalore Project (1987), which had learning activities organized around real-life tasks and focused on the functional use of language.

Although these methods were successful in English language learning, any methodology that proves successful among students may be employed to attain the objective of English language proficiency. As Benzie (2010) points out, 'it is the success with which those skills are applied in the context of use that affects how effectively the language learner can communicate with others' (p. 452).

4 Theoretical Framework

In the current context, the fundamental approach to English language instruction involves immersing students in the language environment. The framework of language immersion (Genesee, 1985; Johnson & Swain, 1997; Johnstone, 2002) is used as a pedagogical approach while engaging students with activities within the class and outside. The idea is to utilize these tasks to consistently immerse ESL learners in their second language to enhance their language proficiency.

The first immersion program was used among Canadian English-speaking students to help them learn French, the other official language of Canada. Genesee (1994) asserts that

> The behaviourist notion of "practice" as a means of learning, which is prevalent in conventional programs, is replaced in immersion-type programs by the notion of "creative construction," in which learners are encouraged to experiment with linguistic forms to communicate with one another and with their teachers about academic and social matters (np).

The immersion method focuses primarily on creating opportunities to interact with each other actively and engaging students in authentic contexts. The notion of grammatical perfection and errors in language use are overlooked to accommodate meaningful communication.

5 Challenges

Similar to teaching every subject, there are some challenges that we face as teachers of English. Some important ones are discussed below:

5.1 Heterogenous Classrooms

Mainly in the Indian context, proficiency in the English language has never been a criterion for admission to an Engineering program. For most students who join the engineering program, L2 (second language) is generally English, L1 being the native tongue. In the Institution where this author teaches, the L2 proficiency of the student group can range from CEFR (Common European Framework of Reference) A2 to C2 as determined by the entry-level test that is mandatory for all students who are admitted into the College B.E., B.Tech., program. This group comprises students from Government rural schools (where the medium of instruction is predominantly the regional language, here Tamil), lateral entry students (that is, students who have completed their diploma and are still coming to grips with the use of the English language), vocational students (who drop the subject like physics in their 11th and 12th grade to take up a vocational course like welding, graphics and multimedia etc., and are similar in the level of English language-use abilities with the lateral entry students), students from the government-aided schools but are from the city (where students do not get many opportunities for English language practice and production), students from the chief minister's 7.5 scheme who are from the state government - aided schools and their ability to use English language is limited and the Prime Minister's Scholarship Scheme (PMSS) students who can hail from any part of the country. The majority of these students fall within the A2- B1 category. Also, there are students from urban areas (who are mainly from the city's elite English medium schools and freely use English outside the classroom). These diverse backgrounds of the students lead to a similar diversity in their English language abilities.

Penny Ur (1999) problematizes the definition of a heterogenous class by stating that there is "no such a thing as a 'homogenous' class since no two learners are similar, and therefore all classes with more than one learner are in fact heterogeneous" (p. 302). Penny Ur (1999) also states that using the term "mixed ability" is misleading because there is a marked difference in the 'ability to perform' of the learners but also their 'potential learning ability'(p. 303).

5.2 *Reluctance to Talk*

Reticence is a significant problem in the English language classroom. It is seen that even the students who are proficient in the language generally are reluctant to participate in language production activities for fear of rejection by their teachers and peers, shame and embarrassment associated with making mistakes, and anxiety of being judged by others, resulting in a lack of self-confidence. One such study by Zhang and Head (2010) among Chinese students who were initially reticent in the speaking class showed significant progress when they were actively involved in the course design and planning of activities, "which would meet their needs and reasons for wanting to speak in English"(p. 1).

5.3 *L1 Interference*

Some students find L1 (their first language or mother tongue) in some way or another interfering in the production of L2. L1 interference may be seen while pronouncing the L2 words (Hago & Khan, 2015). L1 interference may also be seen in speech production, affecting the learners' linguistic (knowledge of language structures such as vocabulary, grammar, and syntax) and communicative competence (Krish & May, 2020). They found a need for more opportunities for practice and limited vocabulary. Studies also show L1 interference with writing proficiency (Chen & Li, 2016).

Understanding the type of interference and its reasons can help overcome the problem. The teacher can provide targeted instruction and foster consistent language practice within and outside the classroom to reduce this interference.

5.4 English as a Foreign Language

Marckwardt (1963) distinguishes English as a Second Language and English as a Foreign Language. Marckwardt (1963) states that English as a Foreign Language refers to English being "taught as a school subject or on an adult level solely to give the student a foreign language competence which he may use in one of several ways to read literature, to read technical works, to listen to the radio, to understand dialogue in the movies, to use the language for communication" (p. 25). When English as a Second Language is used, "the reference is usually to a situation where English becomes a language of instruction in the schools, as in the Philippines, or a lingua franca between speakers of widely diverse languages, as in India" (p. 25).

Although, in the current context, English is taught as a second language, for at least a section of students, English is a foreign language. Due to a lack of English instruction in their schools, these students often find the transition to college overwhelming. Difficulty in understanding and producing the language can be seen at varied levels for individual students. Problems with understanding and speaking in English can be seen during classroom participation, when students comprehend subject lectures, when they communicate with their teachers and peers, and during their day-to-day conversations. Hence, understanding students' diverse needs and enabling language learning is crucial.

5.5 Indifference to Speaking Activities

Speaking as an activity is seldom practised in classrooms. Student feedback indicates that peer interaction between students and teachers is less prevalent in the core engineering departments. This may be due to the teacher's need to maintain classroom decorum, avoid disturbance to neighbouring classes, etc. Hence, when they attend English language classes, students need help engaging in classroom activities, talking, and participating in tasks. In a class where the number of students is around 60 or more, 'happy noise' may be a factor that has to be dealt with. Also, there needs to be more opportunities to test speaking skills in the current tertiary-level curriculum. This, in turn, may lead to lethargy and indifference among students.

5.6 Learner Attitude

Another challenge is the attitude of the learners. Generally, there is a sense of apathy towards English language classes, although they know that learning English is essential. Students may prefer to give importance to their core subjects while neglecting to attend their English classes. Ironically, stress is laid on the students' communicative abilities and sound technical knowledge during placements. Therefore, while enhancing employability skills and communicative proficiency, the English teacher must also promote a congenial learner attitude.

Gardner (1985) underscores the importance of learner attitude:

> ... students learning a second language must be both able and willing to adopt various aspects of behaviour, including verbal behaviour, which characterizes members of the other linguistic-cultural group. This involves cognitive and affective components and emphasizes that cognitive factors such as attitudes and motivation are undoubtedly implicated in second language acquisition. (Gardner, 1985, p. 133)

Research indicates that students' attitudes towards learning English are primarily influenced by the prospects for college admission, career advancement and future job opportunities (Gürsoy, 2013; Johnson, 2001; Wang et al., 2011). Studies also indicate that language learning occurs effortlessly with improved proficiency levels when learners exhibit a positive attitude toward learning (Chamber, 1999; Gardner & Lambert, 1972). Hence, developing a positive attitude towards language learning is vital.

5.7 Learning English as a Subject and not as a Language

Students approach learning the language as they would any other subject. Merely acquiring theoretical competence and rote learning will not suffice while learning a language. Most students say that they aim to be successful in the exam just like in any other subject, like physics or Engineering graphics, and not to learn English as a language. However, others take an interest and practise the language.

6 Strategies to Enhance Language Learning Outcomes

The teacher must create a congenial atmosphere for English language learning and production in the context of all these challenges. Classroom tasks only supplement the primary teaching activity. It may require adjusting teaching strategies according to particular class needs or sometimes even students' needs. Several strategies are adopted to make teaching–learning engaging and enhance the learner experience of the English language in the class. These are some of the teaching strategies that are tried, tested and found valuable and hence adopted:

Two activities implemented experimentally to understand their efficacy will be discussed at length, and other activities tried on a day-to-day basis will be briefly discussed.

6.1 Use of Videos as a Resource

This method has been personally tried, tested, and found to be effective. Integrating videos into English language teaching and learning, particularly videos of TED Talks (Karunakar, 2019), Toastmasters' winning speeches, and other famous speeches like 'I Have a Dream' by Martin Luther King Jr, Charlie Chaplin's 'The Final Speech' from the Great Dictator are influential with learners. It was found that building activities around them encouraged language production across the student groups. Other videos that resonate with learners are those of Dr. Shashi Tharoor, particularly the 'Reparation' speech. This instils confidence among the students when they observe how an Indian and a non-native speaker can hold his own among several native speakers of the language.

Some of the TED Talk videos that were successful with the learners were not by the native speakers of the language, and this diversity of linguistic backgrounds, to a certain extent, puts the student listeners in a comfort zone (Karunakar, 2019). Pranav Mistry's 'Sixth Sense Technology', Matt Cutt's 'Try something new for 30 days', and Muniba Mazari's 'Turning adversity into opportunity' are some examples.

The activities conducted using TED Talks were structured in the following manner:

As a pre-class activity, the videos were put in context before being shown. Some warm-up activities and a quick discussion on the topic were done. An information gap-filling exercise was done as an in-class activity. Post-class activities could include summary writing, etc. Several speaking and writing activities were built around it, and based on Matt Cutt's video, an entire class tried something new for 30 days, resulting in a 2-min presentation and report writing. One student committed to bringing homemade lunch consecutively for 30 days, while another collected used books from friends and family to donate to an orphanage. These additional activities associated with language learning added a refreshing and rewarding aspect to the class, enhancing engagement for all.

Another activity involved using prize-winning Toastmasters speeches. Dhananjaya Hettiarachchi's speech, 'I see something in you, but I don't know what it is' (2014), and Mohammed Qahtani's, 'The power of words' (2015), are very popular with students. Perhaps it is because their accents are closer to home, and hence, students can relate to that.

Video clips of famous movies were also used as in-class material. One practical clip was from the movie 'English Vinglish'; as the tagline suggests, it is about a belittled and insecure Indian housewife enrolled in an English language class. Although it brings back memories of the long-running series, 'Mind your Language' of the 1980s, this movie highlights the difficulties faced by people who cannot communicate in English in an English-speaking country like the U.S. The clip used in the class was the 'Café scene' where the protagonist is ridiculed for her inability to understand and communicate in English. Words like 'Nescafe', popularly pronounced as 'Nescoffee', predominantly has cultural connotations and is associated mainly with south Indians, which the waitress mistakes as 'nice coffee', resonate with the student audience. The choice overload in terms of water (sparkling or still) coffee (Americano, Cappuccino, latte) and fillings (cheese, tomatoes, lettuce) to the point of vexation was another interesting detail that the students enjoyed.

Most of these videos also have a subtitle option, which helps learners with limited English proficiency (LEP) to read and understand the dialogues. Generally, pre-, while and post-viewing activities are given, and students are continuously assessed using classroom observation and language production in classroom activities. The findings show that using videos as a resource in the classroom improved English language production in terms of speaking and encouraged learners to use the language confidently.

6.2 Use of Comics and Graphic Novels as a Resource

Comics and Graphic novels can be used as free and voluntary reading tools in the classroom. Comics is an art form in which all aspects of the narrative are represented by pictorial and linguistic images encapsulated in juxtaposed (placed side by side) panels and pages. Graphic Novels are longer than the typical comic book and most often feature self-contained, rather than continuing stories, with complex plots.

Oriana Gatta (2014) states that despite the pedagogical potential of using comics as educational tools for English language acquisition, little more work needs to be done in this area. This author conducted a study using comics and graphic novels to enhance English language learning among school children (Karunakar, 2017). The study was then used among lower English proficiency learners with appropriate changes at the tertiary level with moderate success. Issa (2017) argues for the potential of comics as effective tools for composition instruction and offers the teaching of English to speakers of other languages and creative writing in undergraduate classrooms using comics. Roozafzai's (2012) study revealed that teaching reading skills using comics is more efficient than teaching them through general texts.

Studies (Krashen & Terrell, 1983; Whitehurst et al., 1988) have shown that the more one reads or is read to, the more it contributes to language improvement. Teachers of English believe that reading fosters improved comprehension, vocabulary development, enhanced writing skills and a more profound mastery of the language; the rationale is that the interplay of pictures and text acts as a scaffold aiding readers to grasp the narrative. Hence, Comics and Graphic Novels, visual and verbal in form, appeal to the students as teaching/learning material.

The activities conducted using Comics and Graphic Novels were structured in the following manner:

Considering the socio-cultural diversity among students in the class and to standardize the reading material, a wide range of books from both Indian and International series were made available. *Spiderman: India* (2004), a comic book series, has a four-part retelling of Spiderman, Pavitr Prabhakar, the namesake and a reimagined version of Peter Parker in a dhoti fighting Nalin Oberoi, the local crime lord of the city of Mumbai. The joint venture between Marvel Enterprises and Gotham Entertainment with authors Jeevan Kang, Suresh Seetharaman, and Sharad Devarajan has successfully reinvented the popular Spiderman series in the Indian environment.

Other series included Amar Chitra Katha, a collection by Editor Anant Pai of India Book House, which presents genuine Indian narratives through retellings of mythology and legend. Tinkle comics, also published by the same company, feature beloved characters such as Akbar & Birbal, Tenali Rama, Supandi, and Shikari Shambu, leaving a lasting impact on readers' minds. In addition, students were also allowed to read any comic books they were interested in. Some of the comics they read included Archie Comics, Calvin and Hobbes, Geronimo Stilton, etc., to name a few.

It is believed that a good reader eventually becomes a good writer. Based on the students' reading activity, two tests were administered to record English language learning. The first one was a gap-filling activity. Sheets with semi-filled word balloons were given, and students were asked to fill them using their imagination and creativity. Words omitted included collocations, subject-verb agreement, if-conditionals, and idioms, which are part of the first-semester 'Technical English' syllabus.

In the second activity, students were prompted to design comic strips using an empty 9-panel grid. They came up with plenty of exciting stories. Their language acquisition was assessed through vocabulary selection, grammar application, coherence in expression, and the narrative's conclusion.

The feedback on both these activities was positive. It is believed that the more one is exposed to the target language, the easier it becomes to master it (Krashen, 1985). While viewing videos, students are constantly immersed in English, including its accent, intonation, pronunciation, and grammar. The comics leverage the picture with the words, making them interesting and meaningful. They also offer language in small chunks, making it easier for the reader to assimilate and register what is being read. The nuanced expressions of the characters aid in comprehending the story, even if readers encounter unfamiliar words due to limited vocabulary.

6.3 *Other Activities in the English Language Class*

Several activities are conducted to improve LSRW skills. Activities that simulate real-life situations in the professional context, like group discussions, interviews and short oral presentations, are conducted to help students during the placement drives. Resource persons from the industry are invited to give a hands-on experience of what is expected of them. Classes in the afternoons and during weekends are especially very challenging. One of the popular activities to break the monotony of the class is using 'Simon Says'.The students, though reluctant initially to participate in this activity, enjoy this exercise of turning around, jumping, sitting, standing, etc., which offers a deviation. A speaking activity like reviewing a movie they have recently watched is also done receptively. Speaking and writing workshops are conducted annually with the department's allocated budget.

So, whether it is watching videos, reading comics or engaging in language learning activities, it is believed that the 'affective filters' (1986), Krashen's term for emotional variables that prevent learning, are said to be in a lowered state. This promotes affability and decreases anxiety and tension, thus providing a conducive atmosphere for English language learning and production.

7 Constructive Feedback to Students

Generally, students are eager to enhance their speaking skills, and after each activity, they request assistance from the teacher. "Ma'am, how was my presentation? Where did I go wrong? How can I improve my English? etc. Teachers may not want to undermine the learner's confidence and correct more. As a general rule, teachers can focus on habitual errors, those that are ingrained in the learners. An example of an oft-committed error is 'can able'—"He can able to do it", or "If they can able to come, they will". Others are 'Cope up, revert back, struck in traffic', substituting 'no' for a question tag, like 'You are coming today, no?', 'Today is hot, no?' etc.

There are various ways to correct errors. One way is to engage in peer correction. The teacher does not need to address a particular student; instead, the teacher should address the whole class or create a task or activity around the errors. Students need to verify the accuracy of spoken language acquired from various sources, as these may not always be reliable. while speaking it could be from their peers, elders, or even teachers. This may lead to easy acceptance of a few wrong usages. Students feel embarrassed when they make mistakes, mainly when speaking. However, students should be taught to embrace their mistakes and practise the skill as that is the only way to learn the language.

8 Limitations

However, there are some limitations to using these methods universally, as what proves effective in one class may yield different results in another. The teacher can foster a supportive and encouraging environment conducive to language learning. However, the students must take responsibility for their learning progression by attending classes, submitting assignments and participating in activities. Also, these methods need to be implemented consistently, with students engaging in immersive learning over some time to determine conclusively whether they are effective.

9 Concluding Remarks

It would be too ambitious to claim that the A2 and B1 level students reached the C1 or C2 level merely due to these interventions. However, it is clear that immersing students in language environments certainly helps overcome L1 interference in L2 acquisition. Incorporating English language learning into the tertiary-level curriculum enhances an engineering graduate's job-preparedness. The importance of English language proficiency in excelling in campus placement interviews, particularly for engineering students, cannot be overstated, as it enables them to clearly articulate their thoughts, ideas, and technical knowledge to potential employers. Furthermore, the current work environment demands collaboration with international clients and teams, making English proficiency an essential asset.

While the dynamic between English and engineering students comes forth as a love-hate relationship, this chapter elucidates that the best approach for teachers of English is to create non-threatening and non-judgemental environments for students to practise their language skills, tailor-make creative and innovative activities for specific classrooms to make language learning interesting, engaging, and effective, particularly in technical academic environments. These learnings are relevant and applicable in all contexts worldwide with similar educational or social situations.

References

Benzie, H. (2010). Graduating as a 'native speaker': International students and English language proficiency in higher education. *Higher Education Research and Development, 29*(4), 447–459. https://doi.org/10.1080/07294361003598824

Chen, Y., & Li, X. (2016). Effects of L1 transfer on English writing of Chinese EFL students. *Cross-Cultural Communication., 12*(4), 50–53.

Chhapia, H. (2023, November 30). Canada, US top 2 study destinations for Indians, U.K. displaces Australia as number 3. *Times of India.* https://timesofindia.indiatimes.com/india/canada-us-top-2-study-destinations-for-indians-uk-displaces-australia-as-no-3/articleshow/105602859.cms

Crystal, D. (2003). *English as a global language.* Cambridge University Press.

Gardner, R. (1985). *Social psychology and second language learning: The role of attitudes and motivation.* Edward Arnold.

Gardner, R., & Lambert, W. (1972). *Attitudes and motivation in second language learning.* Newbury House Publishers.

Gatta, O. (2014). Comic convergence: Toward a prismatic rhetoric for composition studies.

Genesee, F. (1985). Second language learning through immersion: A review of U.S. Programs. *Review of Educational Research, 55*(4), 541–561.

Genesee, F. (1994). Integrating language and content: Lessons from immersion.

Gürsoy, E. (2013). Prospective ELT teacher's attitudes toward the English language in an EFL context. *Journal of International Education Research (JIER), 9*(1), 107–114.

Hago, O. E., & Khan, W. A. (2015). The pronunciation problems faced by Saudi EFL learners at secondary schools. *Education and Linguistics Research, 1*(2), 85–99.

Issa, S. (2017). Comics in the English classroom: A guide to teaching comics across English studies. *Journal of Graphic Novels and Comics, ()*, 1–19. https://doi.org/10.1080/21504857.2017.1355822

Johnson, K. (2001). *An introduction to foreign language learning and teaching.* Pearson Education Limited.

Johnson, R. K., & Swain, M. (1997). *Immersion education: International perspectives.* Cambridge University Press.

Johnstone, R. (2002). *Immersion in a second or additional language at school: A review of the international research.* Scottish CILT.

Karunakar, M. (2017). Using comics and graphic novels as tools to enhance language learning. *The English Classroom., 19*(2), 1–19.

Karunakar, M. (2019). Encouraging English language production using TED talks at the tertiary level: A study in a technical college. *Teaching English with Technology, 19*(4), 57–68.

Kim, M. (2023). A personalized autonomous model to resolve a prolonged dilemma in international students' English language needs in higher education. *Higher Education Research & Development, 42*(3), 603–618.

Kovalyova, Y. Y., Soboleva, A. V., & Kerimkulov, A. T. (2016). Project-based learning in teaching communication skills in English as a foreign language to engineering students. *International Journal of Emerging Technologies in Learning, 11*(4), 153–156. https://doi.org/10.3991/ijet.v11i04.5416

Krashen, S. (1985). The input hypothesis: Issues and implications. Longman.

Krashen, S. D. (1986). *Principles and practice in second language acquisition.* Pergamon Press. http://www.sdkrashen.com/content/books/principles_and_practice.pdf

Krashen, S. D., & Terrell, T. D. (1983). *The natural approach: Language acquisition in the classroom.* Pergamon Press.

Krish, P., & May, O. C. (2020). A case study of L1 interference in speech acts among Chinese L2 students. *3L: Southeast Asian Journal of English Language Studies, 26*(1).

Marckwardt, A. H. (1963). English as a second language and English as a foreign language. *PMLA, 78*(2), 25–28.

National Education Policy. (2020). https://www.education.gov.in/sites/upload_files/mhrd/files/NEP_Final_English_0.pdf

Prabhu, N. S. (1987). *Second language pedagogy.* Oxford University Press.

Roozafzai, Z. S. (2012). The role of comic reading materials in enhancing the ability to read in EFL. *Journal on English Language Teaching, 2*(3), 7–15.

Rus, D. (2020). Creative methodologies in teaching English for engineering students. *Procedia Manufacturing, 46*, 337–343. ISSN 2351-9789. https://doi.org/10.1016/j.promfg.2020.03.049.

Saienko, N., & Shevchenko, M. (2020). Authentic videos in teaching English to engineering students at universities. *International Journal of Learning, Teaching and Educational Research, 19*(8), 350–370.

Salomone, R. (2022). *The rise of English: Global politics and the power of language.* Oxford University Press.

Ur, P. (1999). *A course in language teaching: Practice and theory.* Cambridge Cambridge University Press.

Wang, B. T., Sheu, T. W., Liang, J. C., Tzeng, J. W., & Masatake, N. (2011). The integrated methods of GSP and GSM in concept diagnosis for English grammar. *International Journal of Kansei Information, 2*(2), 87–100.

Whitehurst, G. J., Falco, F. L., Fischel, J. E., DeBaryshe, B. D., Valdez-Menchaga, M. C., & Caulfield, M. (1988). Accelerating language development through picture book reading. *Developmental Psychology, 24*(4), 552–559.

Zhang, X., & Head, K. (2010). Dealing with learner reticence in the speaking class. *ELT Journal, 64*(1), 1–9.

Chapter 3
Intentional Design of Pre-laboratory Modules for First-Year Chemistry Courses

Andrew G. Pearson, Barbara Hadley, Abdullah Karaksha, Brooke E. Harris-Reeves, Kim Cartwright, Mary-Ann Shuker, and Jennifer C. Wilson

Abstract The strategic integration of gamification in Articulate modules has transformed first-year chemistry education and enhanced student learning, engagement, satisfaction, and self-efficacy in laboratories. Chemistry laboratories are crucial for active, experiential learning, fostering skills and problem-solving using scientific methods and equipment. However, students often need help with preparation, leading to stress and confusion. Pre-lab modules were introduced to address this, guiding laboratory concepts and procedures. Continuous feedback from students informed iterative module development, resulting in tailored resources. This approach empowered students with confidence, knowledge, and a sense of ownership over their learning. Embracing diversity, students were given autonomy in their learning journey, with access to multimodal resources, embedded assessments, and immediate feedback. The gamified and adaptive nature of the modules provided a scalable, personalised solution, encouraging the exploration of chemistry concepts. Students reported feeling better prepared and knowledgeable from the modules than traditional reading materials. This student-centric approach reflects our dedication to enhancing the learning experience, nurturing self-efficacy, and fostering active student participation and confidence.

1 Introduction

Traditional teaching methods in first-year chemistry courses typically encompass lectures, textbook readings, problem-solving sessions, and laboratory experiments (Carter & Brickhouse, 1989). Laboratory experiments are an integral part of chemistry education to help students develop essential hands-on skills such as safe handling

A. G. Pearson (✉) · B. Hadley · A. Karaksha · B. E. Harris-Reeves · K. Cartwright · M.-A. Shuker · J. C. Wilson
Griffith University, Gold Coast, QLD, Australia
e-mail: a.pearson@griffith.edu.au

I. Gratchev et al. (eds.), *Transformative Approaches to Career-Ready Education*,
https://doi.org/10.1007/978-981-96-7018-5_3

of chemicals, glassware and equipment usage, and appropriate data collection and interpretation techniques (Veiga et al., 2019). Despite their long-standing use, these methods have several limitations, such as passive learning, memorisation over understanding, limited application of knowledge, and lack of interactivity and technology integration (Karaksha, 2018). Moreover, many students start university studies with limited, if any, laboratory skills. Consequently, students often feel overwhelmed during the lab and do not cope with all the activities necessary to fulfil the learning outcomes (Veiga et al., 2019).

In recent years, this traditional learning approach has increasingly been challenged as the student profile has become more diverse, and innovations in learning and teaching have emerged to address this diversity. The higher education sector has adopted a range of approaches to chemistry education, including active learning strategies (Freeman et al., 2014), flipped classrooms and hybrid teaching (Bokosmaty et al., 2019), and inquiry-based learning to enhance student engagement and prioritise student-centred learning (Kang & Keinonen, 2018). These approaches are designed to promote active participation and foster critical thinking skills essential for chemistry success, encouraging students to engage with the material thoroughly regardless of their backgrounds or abilities.

The COVID-19 pandemic was a significant disruptor to the landscape of higher education, necessitating a sudden pivot from face-to-face to remote/online teaching. This shift accelerated changes in learning experiences for students and led to the rapid adoption of online technology within chemistry education (Fung & Lam, 2020; Schweiker & Levonis, 2020). Students experienced significant challenges that increased disengagement and unpreparedness (Lawrie, 2021). The abrupt transition from in-person to online learning presented numerous challenges for students (DeCoito & Estaiteyeh, 2022). Many struggled to adapt to remote learning platforms, often needing more interactivity and engagement. The absence of face-to-face interaction with peers and academics made it difficult for students to stay motivated and connected to their studies (DeCoito & Estaiteyeh, 2022).

Additionally, technical issues such as connectivity problems, incompatible devices, and unfamiliarity with online platforms could have helped students' ability to engage in coursework (Lawrie, 2021). The social isolation imposed by lockdowns and social distancing measures further impacted students' mental health and well-being (Collantes et al., 2022). The pandemic's economic ramifications and personal and family responsibilities added stress and pressure on many students. Moreover, replicating hands-on laboratory experiences in virtual environments proved challenging (DeCoito & Estaiteyeh, 2022; Huang, 2020). As a result, students may have felt disconnected from the practical aspects of chemistry education, leading to disengagement and complete unpreparedness during laboratory experiments.

In response to student disengagement and unpreparedness during lab sessions, a gamification approach was strategically employed to create a suite of online modules as pre-lab activities. Studies conducted during the COVID-19 pandemic suggested that for online learning to be successful, students must be proactive and self-directed (Mamun et al., 2020). Additionally, incorporating check-in activities has been shown to help students take ownership of their learning process (Ndoye, 2017).

These modules leveraged game design principles and mechanics to enhance student engagement, motivation, and learning outcomes (Karaksha, 2018). This approach offers numerous potential benefits. For example, gamification introduces competition, challenge, and reward elements, making learning more enjoyable and engaging for students (Rudolphi-Solero et al., 2021). By incorporating interactive elements such as quizzes and virtual labs, students were motivated to actively participate and explore chemistry concepts in a fun and immersive manner. Gamification provides intrinsic motivation by aligning with students' desires for achievement, progression, and mastery (Wu et al., 2020). Moreover, gamified online modules encourage active learning by requiring students to solve problems, make decisions, and apply concepts in simulated real-world scenarios. This hands-on approach promotes more profound understanding and retention of chemistry principles than passive learning methods such as lectures or readings (Karaksha, 2018; Nieminen et al., 2021). Gamification enables real-time feedback and assessment mechanisms that provide immediate insight into students' understanding and performance. By offering constructive feedback, hints, and guidance throughout the learning process, online modules help students identify areas for improvement and take proactive steps to strengthen their knowledge and skills (Ravyse et al., 2017). Finally, gamified online modules offer greater accessibility and flexibility than traditional classroom-based instruction. Students can access gamified learning materials anytime, anywhere, allowing for self-paced learning and accommodating diverse schedules and learning styles (Krishnamurthy et al., 2022; Larson, 2020).

These teaching innovations equip students with indispensable skills and competencies vital for today's dynamic job market. Beyond enriching academic knowledge, these approaches foster adaptability, problem-solving prowess, and technological fluency, highly sought after by employers across diverse industries.

2 Method

2.1 Context for Module Development

The initial iterative design process was based on cognitive and pedagogical theories, ensuring that the material was engaging and pedagogically sound, ensuring effective achievement of learning outcomes. The development was based on principles derived from Bloom's Taxonomy and scaffolded learning strategies (Athanassiou et al., 2003), emphasising the importance of building upon learners' existing knowledge and progressively increasing the complexity of added information. This information was already available to us from previous and parallel courses.

2.2 Module Design and Development

The first step in the content design involved the identification of learning objectives to ensure they aligned closely with both the existing curriculum and the needs of the students. It was essential to consider the student's pre-existing knowledge and keep the technological aspects simple to increase their learning experiences without requiring them to gain new technological skills. The Articulate platform was chosen as the best way to cater to this requirement by offering modules accessible across a broad spectrum of internet connectivity options—from interactive online modules to PDF versions for offline access—and device types, ranging from desktop computers to tablets and mobile phones.

The development of each module adhered to the Web Content Accessibility Guidelines (World Wide Web Consortium, 2008), integrating features such as alternative text for images, video captions, and seamless navigational structures to enhance accessibility for students with disabilities. The Universal Design for Learning (UDL) principles (CAST, 2018) were followed to ensure a multifaceted approach to presenting information and engaging students. This encompassed various strategies, including narrating extensive text blocks and accommodating diverse learning preferences and needs.

The design and development phase of both the content and lab modules used storyboard techniques to map out each module's content flow, visual elements, and interactive components (Orr et al., 1994). Following the categorisation of objectives under thematic headings, the content was interspersed with multimedia elements like laboratory videos and chemical reaction simulations, which have proven especially effective for teaching chemistry concepts. These thematic headings facilitated the revision of the modules and allowed the students to find the sections they wished to revisit quickly.

The content compilation was from various sources, such as the designated textbook, lecture slides, research articles, laboratory manuals, and insights from previously conducted workshops. This ensured comprehensive coverage of all learning objectives. To reinforce understanding, the same concept was presented from multiple optional perspectives to aid students in grasping the more challenging principles—for example, the theory of acids and bases. The first explanation was purely scientific, where acids and bases interact to form water and salts. The second explanation defined the acids and bases as the donation and acceptance of protons using the sour taste of lemons and the bitter taste of baking soda as the acid and base. The third explanation was an analogy with children on a seesaw balancing either side with protons to show neutrality at different pH values. Traditional glossaries were replaced with hyperlinks to definitions and further information to enhance learning. Additionally, optional links to a wide range of informational resources were strategically placed throughout the module. This approach offers learners a layered learning experience, ranging from essential remedial content to more extensive exploration.

Integrating online pre-laboratory modules into the chemistry curriculum has changed how students approach and understand chemical experiments.

Each module contained:

- Theory
- Material and Equipment
- Experimental Procedure
- Knowledge Check
- Feedback Survey

Within each module, there are interactive elements that include instructional videos, external links to videos, hyperlinks, learning objectives, relevant quotes from famous scientists, multiple-choice questions, fill-in-the-blank questions, sorting games using cards that drag from one area to another, flip cards, mini-quizzes, layered tabs that they click to open, interactive timelines to explore, and downloadable worksheets. These diverse resources enhanced the students' engagement and understanding of the subject matter.

Figure 1 shows an example from a module where the units and abbreviations must be matched. Employing an interactive approach, including matching names to abbreviations rather than passive reading, enhanced learning engagement and retention. This method encourages active participation, which helps students internalise and recall information more effectively, making it a superior strategy for mastering complex material. For example, the activity shown in Fig. 1 helps students link the μ symbol, which is new to many students, to the term "micro".

Figure 2 is an example of a Storyline360 build where students must input the correct number of significant figures in a number. Incorporating a virtual online calculator where students input data and receive immediate feedback further enriches the learning experience. If the input is correct, a character cheers, providing positive reinforcement that boosts student morale. Conversely, if the answer is incorrect, students receive constructive feedback designed to guide them gently toward the correct solution, allowing unlimited attempts. This approach fosters a supportive learning environment, encourages perseverance, and deepens understanding through

Fig. 1 Unit matching activity from measurements pre-laboratory module

Fig. 2 Calculator activity for significant figures

trial and improvement. Using the calculator with constructive feedback and non-judgement can reduce the stress and stigma associated with making errors, normalising learning through failure while making learning more engaging and less intimidating. Additionally, this approach provides instant feedback, crucial for learning complex concepts such as significant figures. Instead of simply marking an answer wrong, the calculator can guide students towards understanding why their response was incorrect, helping them to grasp the underlying principles more effectively. This provides the equivalent of a personal tutor for every student when and where they need them. This method also allows students unlimited attempts, promoting a mastery-based learning environment where students can learn at their own pace and ensure they fully understand the concept before moving on.

Figure 3 shows the use of questions on flip cards, which allows the students to actively engage with the content enhancing memory retention and making learning more interactive compared to passive text reading.

Figure 4 shows an example of a mathematical equation question. Even though the answer is in digits, the calculations behind the answer are quite complex. Hence, some students can adeptly answer the question and input the correct answer, whereas others need help and answer it incorrectly. When this occurs, the module allows them to see the answer explained by viewing the solution process or skip to the next question. This choice respects their judgment about their learning and enhances their engagement by allowing them to decide how deeply to delve into a topic. If the student has answered it correctly, watching an explanation of a concept they already understand can be frustrating. The feedback from students was that they appreciated this flexibility, as it enabled them to reinforce their understanding when needed or progress faster if they felt confident, fostering a sense of autonomy and personal investment in their

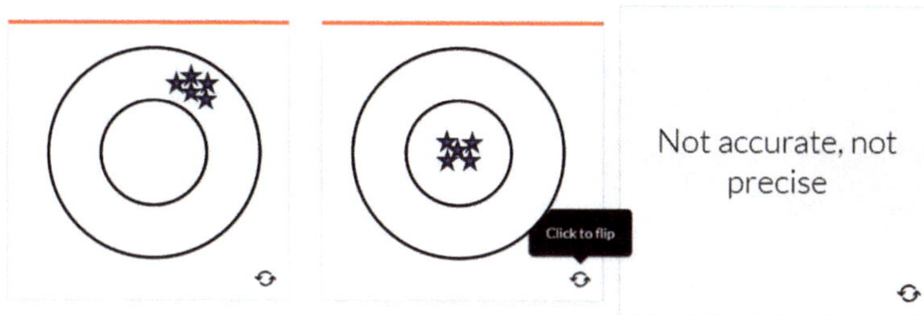

Fig. 3 Flip cards that show images and/or text on either side of the card

educational journey. This allowed the students to tailor their learning, consistent with Pink's motivation theory, which emphasises autonomy, mastery, and purpose as key factors in motivating individuals (Pink, 2009).

Figure 5 shows a still from a video lab demonstration. Having students watch a video demonstration of a scientific lab experiment before conducting it themselves offers numerous educational benefits. It provides a clear, visual understanding of

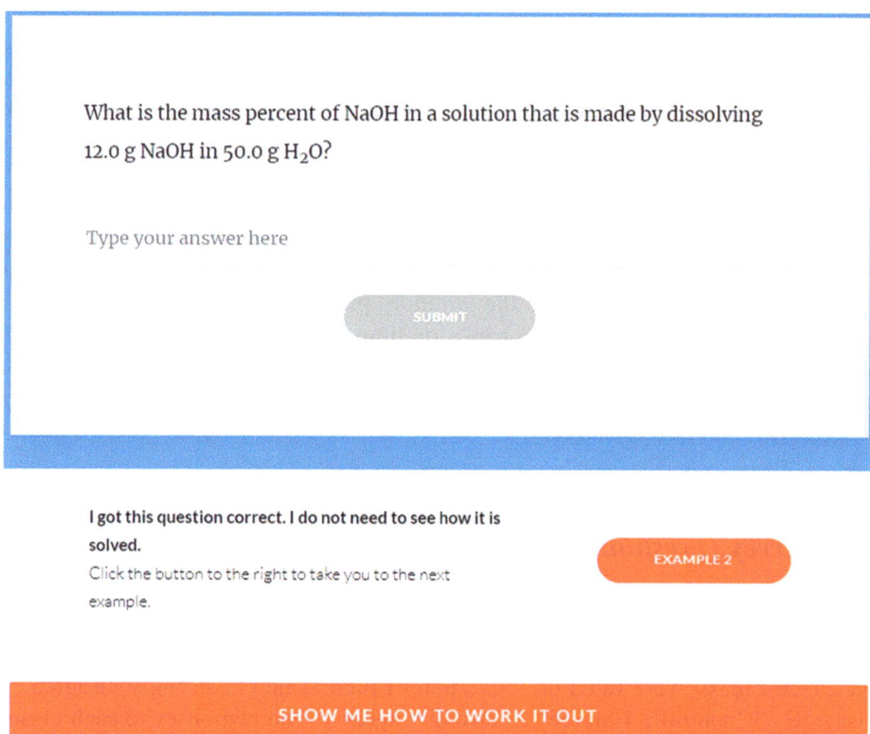

Fig. 4 Fill in the blank activity, followed by the option to skip or view the working out

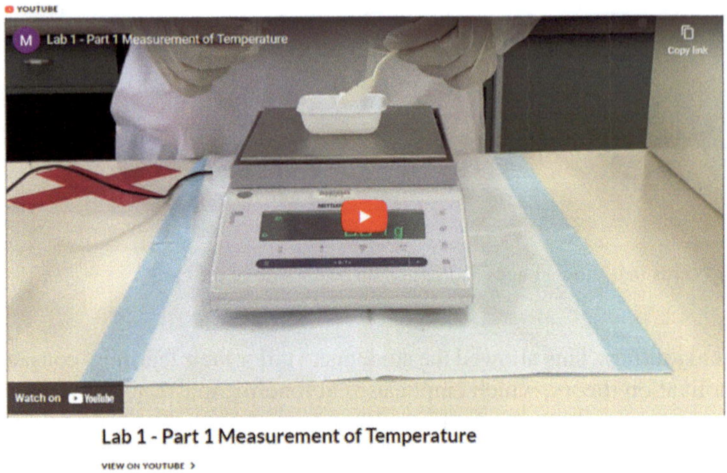

Fig. 5 Screenshot of a video of the demonstration of laboratory technique

the procedures and techniques involved, helping to clarify any complexities of the experiment which might be difficult to grasp through text alone. This prior exposure reduces ambiguity and increases student confidence as they replicate the procedures. Moreover, watching a demonstration allows students to observe proper safety protocols and equipment handling, which is crucial for maintaining a safe laboratory environment. This pre-emptive visual guide also helps minimise errors during the experiment, as students are more likely to remember the steps and techniques they have seen performed. Additionally, pre-lab video demonstrations can enhance lab time efficiency. Students come to the lab more prepared and have a better idea of the experiment's flow, maximising hands-on time by reducing the need for lengthy explanations and corrections during the lab session. This preparatory tool supports various preferred learning modes, particularly benefiting visual learners who might struggle with text-based instructions.

2.3 Survey Questions

Student perceptions of the pre-laboratory modules were collected using purpose-designed surveys that were integrated into the end of each module. Questions 1 and 2 (see Fig. 6) were rated on a five-point Likert scale (1 'strongly disagree', 2 'disagree', 3 'neutral', 4 'agree', 5 'strongly agree'). The responses to each closed Likert statement were averaged, and variation was represented as standard deviation from the mean. Questions 3–5 (see Tables 2, 3 and 4) were open-ended and sought feedback on what was positive and negative about the modules and what could be

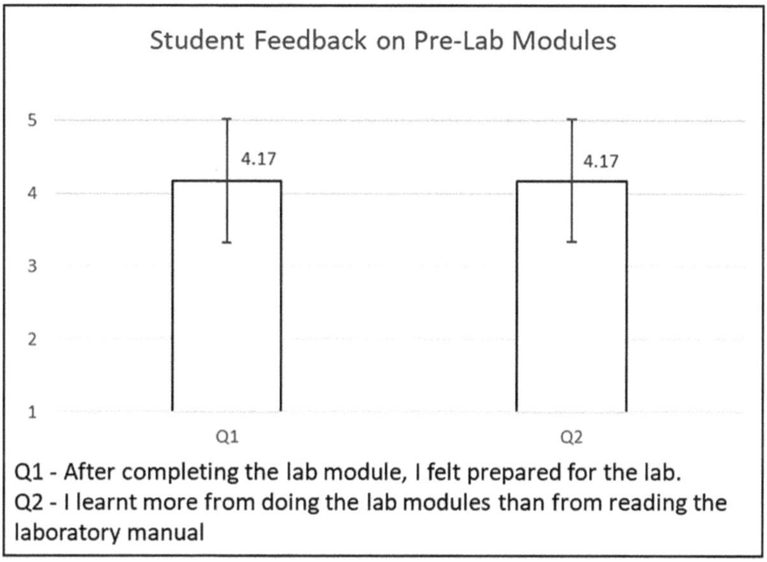

Fig. 6 Student perceptions of the pre-lab modules

improved. The participant feedback obtained from the three open-ended questions in the post-survey was systematically analysed by categorising responses into distinct categories, as described below in the data analysis and reported as frequency counts and percentages.

2.4 Data Analysis, Feedback and Improvements

This study employed a mixed-methods approach, incorporating qualitative and quantitative data collection methods in the surveys. Quantitative data was reported using calculated mean and standard deviation values.

A reflexive qualitative analysis approach was employed (Braun & Clarke, 2006) to examine the responses provided by participants regarding their experiences with the online lab modules. This involved identifying recurring patterns or themes within the data, which gave a deep understanding of participants' perspectives and experiences. The qualitative data was first thoroughly reviewed to familiarise with the content and understand the range of responses. Each response was systematically analysed, and relevant text segments were identified and coded based on their content. Codes were typically short descriptive labels that captured the essence of the response. For example, responses expressing positive aspects like "clear explanation" or "video demonstrations" were coded accordingly. As coding progressed, similar codes were grouped to form preliminary themes. Codes representing similar concepts or ideas were clustered to identify broader data patterns. The preliminary themes were further refined and developed through iterative analysis. Themes emerged as patterns across multiple responses, highlighting commonalities in participants' feedback. The identified themes were continuously reviewed and refined to ensure they accurately captured the essence of the data. Any discrepancies or uncertainties in coding were addressed through discussion and consensus among the researchers. Once the themes were finalised, they were interpreted in the context of the research objectives and existing literature. The significance of each theme was considered in relation to the overall research question and its implications for improving the online lab modules. Finally, the identified themes and supporting quotes were reported in a structured manner, providing a clear and comprehensive overview of the qualitative findings.

A notable advantage of preceding traditional coding in favour of a reflexive thematic analysis was the efficiency of time and resource allocation. Furthermore, the flexibility of this method allowed for a more intrinsic identification of patterns deeply rooted in the team's extensive engagement with the data. Given the team's role in authoring the modules, this approach also enabled a deeper understanding of the module's complexities. Despite the inherent subjectivity of this analytical method, it was deemed an asset, prompting iterative enhancements to the modules to improve student interaction and clarify essential concepts. This methodology proved particularly effective for exploratory studies, with goals encompassing upkeep (maintenance of the modules), statistical scrutiny, error detection, and aesthetic refinements.

Over the past four years, we have meticulously refined our online laboratory modules, incorporating feedback from all respondents to enhance functionality and design. This process involved collecting and analysing student responses to pinpoint areas for improvement as outlined previously, ensuring that each version of the modules rectified previous errors and aligned more closely with the users' diverse needs and aesthetic preferences. The culmination of this extensive feedback loop has resulted in a series of final products that are error-free, visually appealing, intuitively designed, and support many diverse ways of learning. These redesigned modules offer a seamless and engaging learning experience, demonstrating our commitment to educational excellence and responsiveness to learner feedback. Through this approach, we have successfully transformed initial concepts into polished educational tools that effectively support and enrich the learning journey.

2.5 *Human Research Ethics*

Student demographic data is reported from the Griffith University statistics portal for comparison. This ensured that each participant was not required to provide detailed demographic data to the research team. This intentional omission from the surveys ensured compliance with human research ethics guidelines and optimised de-identification of personal information, further safeguarding participant confidentiality. These requirements ensured compliance with the Griffith University Human Research Ethics Committee approval for the project (Ethics no. 2019/398).

Table 1 Demographic data for participants

Status	Commencing	97.8%
	Continuing	2.2%
	Domestic	86.3%
	International	13.7%
	First In Family	40.3%
	Disability	6.7%
Gender	Male	36.9%
	Female	62.8%
Age (years)	17–19	66.0%
	20–24	19.9%
	25–29	7.8%
	30–39	5.5%
	40–49	0.9%
Bachelor of	Biomedical Science	20%
	Pharmacy	17%
	Health Science	16%
	Exercise Science	9%
	Dental Health Science	9%
	Nutrition and Dietetics	7%
	Medical Science	6%
	Pharmacology and Toxicology	4%

3 Results

3.1 Demographic Data

Participants were enrolled in Chemistry of Biological Systems I and/or Chemistry of Biological Systems II at Griffith University, Australia, between 2020–2023. The surveys were anonymous, and no demographic data was collected. The institutional student profile indicated that the 3013 students had the demographics shown in Table 1.

3.2 Quantitative Survey Data Results

Figure 6 displays the aggregated responses from post-module surveys administered to students to solicit feedback regarding their perceptions of the modules. The survey questions were designed to probe whether students felt adequately prepared for laboratory work upon completion of the module (Q1) and whether they perceived a greater

Table 2 Survey question—What was positive about this online lab module?

Category	Frequency count	Frequency percentage
Videos and Educational Content	579	43.8%
Interactive	303	22.9%
Clarity and Conciseness	106	8.0%
Visual Aids	87	6.6%
Engagement	72	5.5%
Ease of Understanding	59	4.5%
Instruction Quality	48	3.6%
Real-life Examples	22	1.7%
Comprehensive Coverage	15	1.1%
Practical Application	13	1.0%
Ease of Access	11	0.8%
Lab Preparation	6	0.5%

Table 3 Survey question—"What was negative about this online lab module?"

Category	Frequency count	Frequency percentage
Confusion	25	65.8%
Content	7	18.4%
Engagement	5	13.2%
Technical Issues	1	2.6%

Table 4 Survey question—"What suggestions do you have for improvement of this online lab module?"

Category	Frequency count	Frequency percentage
Increase Engagement	33	66.0%
Enhancing Clarity	14	28.0%
Adjust Content	3	6.0%

knowledge acquisition from the module compared to solely reading the laboratory manual (Q2). The findings suggest that, on average, participants concurred that the online lab modules adequately prepared them for laboratory activities and garnered more insights from the modules than from reading the laboratory manual alone. The elevated mean scores, approaching four on the Likert scale denoting "Agree," coupled with a substantial response rate, indicate a favourable reception of the online lab modules. Furthermore, the minimal standard deviations imply a narrow dispersion of responses, signifying a consensus among participants.

3.3 Qualitative Survey Data Results

The impact of online laboratory modules on students' knowledge acquisition supports their perception that the module design was structured to provide a diverse range of engagement modalities (see Table 2). Specifically, one of the questions encouraged students to reflect on the favourable aspects of the modules. A thematic analysis of these positive aspects revealed that *'videos and educational content'* emerged as the most frequently cited response ($N = 579$). For example, students commented, "*I acquired concepts through a combination of videos, textual material, and quizzes*", and "*The procedure videos helped get an idea about what exactly we will be doing in the lab and feeling prepared for the lab*", and "*I liked the videos and easy instructions—not too much info*". Further categories identified through this analysis, which support the positive design aspect of the module, include: *'interactive'* ($n = 303$), indicating mentions of interactivity or interactive elements within the modules; *'visual aids'* ($n = 87$), encompassing references to diagrams, animations, or other visual aids aside from videos; *'engagement'* ($n = 72$), denoting references to the material being engaging or interesting; and *'instruction quality'* ($n = 48$). Notably, the feedback received encompassed more than just the design elements of the module. Further, the analysis indicated that students valued the module's '*clarity and conciseness'* ($n = 106$), its '*ease of understanding'* ($n = 59$), and the perceived '*quality instruction'* ($n = 48$).

In the analysis of student feedback regarding the modules' positive and negative aspects, it was evident that the responses emphasising the positive aspects (2341 comments) were notably more abundant than those highlighting negative aspects (552 comments). Many respondents expressed no negative sentiments towards the modules (524 comments such as "–", "n/a", "nothing", and "nil"). However, examining the limited data about negative feedback about the online lab modules revealed discernible categories, shedding light on various aspects perceived negatively by respondents, encompassing issues such as confusion, technical difficulties, content redundancy, and lack of engagement, as shown in Table 3. Notably, the most frequently cited category was *'confusion'* ($n = 25$), suggesting that enhancing clarity and providing additional examples could enhance comprehension. Student testimonies reflect this sentiment: "*Certain concepts were challenging to grasp online and required more illustrative examples. Understanding the vernier calliper was particularly perplexing in the online format. Additionally, clarifying whether the quiz contributes to our overall grade would be beneficial*" and "*It is hard to visualise the equipment, like the pipette*".

The category of negative comments that garnered the second highest frequency of mentions was *'content'* ($n = 7$). Despite its low occurrence, exploring student feedback concerning their perceptions of module content is imperative to foster a positive learning experience. Students found the absence of interactive elements problematic for addressing doubts; they felt overwhelmed by the volume and complexity of information, perceived some content as repetitive or irrelevant, and needed clarification about the assessment methods and how the online modules would impact their overall understanding and grades. Early remarks from students expressing dissatisfaction with the content aspect of the module include statements such as: "*The content appears lengthy, and revisiting certain sections felt tedious due to familiarity*" and "*Certain online materials were challenging to grasp without adequate illustrative examples.*" The length of the modules was subsequently addressed by allowing students to skip familiar concepts.

In the thematic analysis of responses to the query "*What suggestions do you have for improvement of this online lab module?*", a substantial portion of respondents expressed satisfaction, indicating no perceived need for enhancements. However, upon closer examination of the data where suggestions were provided, three distinct categories emerged, as shown in Table 4, shedding light on areas where respondents identified potential areas for improvement within the online lab modules. The foremost category cited most frequently emphasised the need to *'increase engagement'* ($n = 33$). This recommendation implied a desire for heightened interactivity within the content, such as incorporating periodic quizzes. Testimonials from students underscored this sentiment, with one stating, "*more interactive content, e.g., small quizzes throughout.*"

The second most prevalent category was related to *'enhancing clarity'* ($n = 14$), primarily by including additional examples. Respondents indicated a preference for clearer explanations to facilitate comprehension. Testimonials supporting this notion included suggestions such as, "*while maintaining content integrity, consider condensing it; additionally, allow users to navigate non-sequentially,*" and "*perhaps incorporate more examples illustrating the use of a Vernier calliper.*"

4 Discussion

The survey results indicate positive perceptions of the online pre-lab modules among students. Both questions regarding the adequacy of preparation for laboratory work and the perceived acquisition of knowledge compared to reading a lab manual received high mean scores, with both at 4.17 on a Likert scale of 5.

The thematic analysis of positive feedback highlighted several critical aspects of the module design that students found particularly beneficial. These include incorporating videos and educational content, interactivity, visual aids, engagement, and the quality of instruction. Notably, students appreciated the clarity, conciseness, and ease of understanding of the modules, as well as the inclusion of real-life examples and comprehensive coverage of topics. These findings align with previous research emphasising the importance of multimedia content, interactivity, and explicit instruction in online learning environments (Domagk et al., 2010; Zhang et al., 2006).

The student feedback also revealed areas where improvements could be made. The most common suggestion for enhancement was to increase engagement through more interactive content, such as periodic quizzes. Some students desired more precise explanations and additional examples to aid comprehension. Additionally, some students raised concerns about confusion, technical issues, content redundancy, and lack of engagement. These areas were addressed in an iterative approach to improve the student experience. These findings underscore the importance of addressing these issues to optimise the effectiveness of online laboratory modules.

These findings are consistent with existing literature suggesting that well-designed online modules can effectively prepare students for laboratory activities and enhance knowledge acquisition compared to traditional methods such as reading laboratory manuals alone (Jolley et al., 2016).

The results offer valuable insights that can inform the design of online learning environments in disciplines beyond science and laboratory-based courses. The design of further modules will emphasise the following:

- Multimedia Integration
  - Incorporating multimedia elements such as videos, animations, and visual aids can enhance learning experiences across disciplines. Multimedia can provide alternative means of presenting complex concepts, engaging students, and catering to diverse learning preferences.
- Interactivity and Engagement
  - The emphasis on interactivity and engagement within the online modules highlights the importance of active learning.
- Clarity and Conciseness
  - Clear and concise instruction is essential for effective learning in any subject area. Designing online modules with clear learning objectives, organised content, and straightforward explanations can improve comprehension and retention of information.
- Real-life Examples and Application
  - Incorporating real-life examples and practical applications can help students connect theoretical concepts to real-world contexts.
- Assessment and Feedback
  - Regular feedback and assessment opportunities are crucial for monitoring student progress and promoting continuous learning. In online learning environments across disciplines, incorporating quizzes, assignments, peer assessments, and feedback mechanisms can help assess student understanding and provide timely guidance for improvement.
- Addressing Technical Issues and Confusion
  - Ensuring technical reliability and addressing confusion or misconceptions promptly are critical aspects of online learning design. Clear communication channels, technical support resources, and proactive troubleshooting strategies can help mitigate technical issues and enhance the overall learning experience in any discipline.

- Customisation and Adaptability
  - Designing online learning environments that allow for customisation and adaptability can accommodate diverse learner needs and preferences. Whether offering flexible pacing options, adaptive learning pathways, or personalised content recommendations, customisation features can optimise learning outcomes.
- Continuous Improvement and Feedback Loop
  - Establishing mechanisms for collecting student feedback and continuously improving online learning environments is essential. Regular evaluations, surveys, and focus groups can provide valuable insights into student experiences and inform iterative design enhancements in any discipline.

5 Conclusion

In conclusion, the results suggest that the online laboratory modules were well-received by students and effectively facilitated their preparation for laboratory work and knowledge acquisition. Thematic analysis of student feedback highlighted students' appreciation for many aspects of module design, including multimedia integration, interactivity, clarity, and comprehensive coverage. These findings align with previous research underscoring the importance of multimedia content and explicit instruction in online learning.

Through continuous cycles of deployment and feedback, the modules were improved in areas such as increasing engagement, enhancing clarity, and addressing technical issues and confusion. This enhanced the effectiveness of online laboratory modules and improved the overall learning experience for students. The study underscores the significance of iterative design enhancements informed by student feedback to create engaging and effective online learning environments.

As the education landscape evolves, incorporating innovative teaching methods like online laboratory modules is essential for preparing students to enter the workforce with the necessary skills and knowledge. By addressing student feedback and continuously refining these modules, educators can ensure that graduates are academically proficient and equipped with the practical abilities and adaptability required in their future careers.

References

Athanassiou, N., McNett, J. M., & Harvey, C. (2003). Critical thinking in the management classroom: Bloom's taxonomy as a learning tool. *Journal of Management Education, 27*(5), 533–555. https://doi.org/10.1177/1052562903252515

Bokosmaty, R., Bridgeman, A., & Muir, M. (2019). Using a partially flipped learning model to teach first-year undergraduate chemistry. *Journal of Chemical Education, 96*(4), 629–639. https://doi.org/10.1021/acs.jchemed.8b00414

Braun, V., & Clarke, V. (2006). Using thematic analysis in psychology. *Qualitative Research in Psychology, 3*(2), 77–101. https://doi.org/10.1191/1478088706qp063oa

Carter, C. S., & Brickhouse, N. W. (1989). What makes chemistry difficult? Alternate perceptions. *Journal of Chemical Education, 66*(3), 223. https://doi.org/10.1021/ed066p223

CAST. (2018). Universal design for learning guidelines version 2.2. Retrieved from http://udlguidelines.cast.org

Collantes, L. M., Torres, J. M., Astrero, E. T., Gaboy, R. G., Castillo, M. E. G. C., & Mukminin, A. (2022). Perspectives, challenges, and opportunities: The pandemic teaching experiences in science courses. *Journal of Higher Education Theory and Practice, 22*(4). https://doi.org/10.33423/jhetp.v22i4.5131

DeCoito, I., & Estaiteyeh, M. (2022). Transitioning to online teaching during the COVID-19 pandemic: An exploration of STEM teachers' views, successes, and challenges. *Journal of Science Education and Technology, 31*(3), 340–356. https://doi.org/10.1007/s10956-022-09958-z

Domagk, S., Schwartz, R. N., & Plass, J. L. (2010). Interactivity in multimedia learning: An integrated model. *Computers in Human Behavior, 26*(5), 1024–1033. https://doi.org/10.1016/j.chb.2010.03.003

Freeman, S., Eddy, S. L., McDonough, M., Smith, M. K., Okoroafor, N., Jordt, H., & Wenderoth, M. P. (2014). Active learning increases student performance in science, engineering, and mathematics. *Proceedings of the National Academy of Sciences, 111*(23), 8410–8415. https://doi.org/10.1073/pnas.1319030111

Fung, F. M., & Lam, Y. (2020). How COVID-19 disrupted our "flipped" freshman organic chemistry course: Insights gained from Singapore. *Journal of Chemical Education, 97*(9), 2573–2580. https://doi.org/10.1021/acs.jchemed.0c00590

Huang, J. (2020). Successes and challenges: Online teaching and learning of chemistry in higher education in China in the time of COVID-19. *Journal of Chemical Education, 97*(9), 2810–2814. https://doi.org/10.1021/acs.jchemed.0c00671

Jolley, D. F., Wilson, S. R., Kelso, C., O'Brien, G., & Mason, C. E. (2016). Analytical thinking, analytical action: Using prelab video demonstrations and e-quizzes to improve undergraduate preparedness for analytical chemistry practical classes. *Journal of Chemical Education, 93*(11), 1855–1862. https://doi.org/10.1021/acs.jchemed.6b00266

Kang, J., & Keinonen, T. (2018). The effect of student-centred approaches on students' interest and achievement in science: Relevant topic-based, open and guided inquiry-based, and discussion-based approaches. *Research in Science Education, 48*, 865–885. https://doi.org/10.1007/s11165-016-9590-2

Karaksha, A. (2018). The scholarship of learning and teaching in the dynamic discipline of pharmacology and Chemistry. In *Emerging technologies and work-integrated learning experiences in allied health education* (pp. 12–36). IGI Global. https://doi.org/10.4018/978-1-5225-3850-9.ch002

Krishnamurthy, K., Selvaraj, N., Gupta, P., Cyriac, B., Dhurairaj, P., Abdullah, A., Krishnapillai, A., Lugova, H., Haque, M., Xie, S., & Ang, E. T. (2022). Benefits of gamification in medical education. *Clinical Anatomy, 35*(6), 795–807. https://doi.org/10.1002/ca.23916

Larson, K. (2020). Serious games and gamification in the corporate training environment: A literature review. *TechTrends, 64*(2), 319–328. https://doi.org/10.1007/s11528-019-00446-7

Lawrie, G. (2021). Chemistry education research and practice in diverse online learning environments: Resilience, complexity and opportunity! *Chemistry Education Research and Practice, 22*(1), 7–11. https://doi.org/10.1039/d0rp90013c

Mamun, M. A. A., Lawrie, G., & Wright, T. (2020). Instructional design of scaffolded online learning modules for self-directed and inquiry-based learning environments. *Computers & Education, 144*, 103695. https://doi.org/10.1016/j.compedu.2019.103695

Ndoye, A. (2017). Peer/Self assessment and student learning. *International Journal of Teaching and Learning in Higher Education, 29*(2), 255–269.

Nieminen, J. H., Asikainen, H., & Rämö, J. (2021). Promoting deep approach to learning and self-efficacy by changing the purpose of self-assessment: A comparison of summative and formative models. *Studies in Higher Education, 46*(7), 1296–1311. https://doi.org/10.1080/03075079.2019.1688282

Orr, K. L., Golas, K. C., & Yao, K. (1994). Storyboard development for interactive multimedia training. *Journal of Interactive Instruction Development, 6*(3), 18–31.

Pink, D. H. (2009). *Drive: The surprising truth about what motivates Us.* Riverhead Books.

Ravyse, W. S., Seugnet Blignaut, A., Leendertz, V., & Woolner, A. (2017). Success factors for serious games to enhance learning: A systematic review. *Virtual Reality, 21*, 31–58. https://doi.org/10.1007/s10055-016-0298-4

Rudolphi-Solero, T., Jimenez-Zayas, A., Lorenzo-Alvarez, R., Domínguez-Pinos, D., Ruiz-Gomez, M. J., & Sendra-Portero, F. (2021). A team-based competition for undergraduate medical students to learn radiology within the virtual world of second life. *Insights into Imaging, 12*, 1–12. https://doi.org/10.1186/s13244-021-01032-3

Schweiker, S. S., & Levonis, S. M. (2020). Insights gained while teaching first semester chemistry in the time of COVID-19 at Bond University in Australia. *Journal of Chemical Education, 97*(9), 2863–2865. https://doi.org/10.1021/acs.jchemed.0c00621

Veiga, N., Luzardo, F., Irving, K., Rodríguez-Ayán, M. N., & Torres, J. (2019). Online pre-laboratory tools for first-year undergraduate chemistry course in Uruguay: Student preferences and implications on student performance. *Chemistry Education Research and Practice, 20*(1), 229–245. https://doi.org/10.1039/c8rp00204e

World Wide Web Consortium. (2008). Web Content Accessibility Guidelines (WCAG) 2.0.

Wu, H., Li, S., Zheng, J., & Guo, J. (2020). Medical students' motivation and academic performance: The mediating roles of self-efficacy and learning engagement. *Medical Education Online, 25*(1), 1742964. https://doi.org/10.1080/10872981.2020.1742964

Zhang, D., Zhou, L., Briggs, R. O., & Nunamaker, J. F., Jr. (2006). Instructional video in e-learning: Assessing the impact of interactive video on learning effectiveness. *Information & Management, 43*(1), 15–27. https://doi.org/10.1016/j.im.2005.01.004

Chapter 4
Engineering Problem Solving

Shreepad Karmalkar and Srikant Gollapudi

Abstract The ultimate goal of engineering is products and processes aimed at practical ends in real life. After reading this material, the readers will be able to explain this goal, the nature of engineering, the knowledge and skills required in engineering problem-solving, and the discipline-independent strategies employed to solve a general problem. First, we discuss discipline-independent general thinking strategies and barriers to problem-solving with examples. The strategies discussed include—trial and error, logical reasoning, divide and conquer, representation/modelling, analysis, reduction, reformulation and analogy. The three steps involved in the analysis of any problem are highlighted, namely—separation of the whole into parts, understanding the parts in isolation, and combining the understanding so obtained to understand the whole. Two barriers to problem-solving, namely—mental set and functional fixedness, are discussed. Next, we identify three levels of engineering, namely—software, hardware systems and components, and highlight the interdisciplinary nature of engineering problems by considering two real-life examples. One example is developing a fabrication process for a very high-frequency varactor tuning diode with a high-quality factor. Another example is the development of earthquake-resistant drywalls. We discuss the solutions to these problems, highlighting how general thinking strategies with discipline-specific concepts are employed in the solution. Further, this discussion shows how engineering problem-solving requires identifying variables that can be ignored in a given situation based on hands-on experience and intuition, critical thinking, creative thinking and communication skills.

S. Karmalkar (✉)
School of Electrical and Computer Sciences, Indian Institute of Technology Bhubaneswar, Bhubaneswar, India
e-mail: karmal@iitbbs.ac.in

S. Gollapudi
School of Minerals, Metallurgical and Materials Engineering, Indian Institute of Technology Bhubaneswar, Bhubaneswar, India

I. Gratchev et al. (eds.), *Transformative Approaches to Career-Ready Education*,
https://doi.org/10.1007/978-981-96-7018-5_4

1 Engineering

The ultimate goal of engineering is the development of products and processes aimed at practical ends in real life. Hence, the solutions to most engineering problems are interdisciplinary, requiring a convergence of science, mathematics, logic, economics, and ethics. Furthermore, such engineering solutions are appreciated and adopted as long as they are practical and economical. In this chapter, we discuss different strategies for solving engineering problems. There are three levels of engineering: software, hardware systems, and components. Abacus and Matlab are examples of software.

On the other hand, supercomputers, satellites and integrated circuits are examples of systems. Parts of a system are called components. Transistors, resistors, and capacitors are components that constitute many electronic systems.

We shall illustrate the issues involved in engineering problem-solving by considering two real-life examples. One example is component-level engineering. It involves developing a fabrication process for a very high-frequency varactor tuning diode with a high-quality factor. Another example is system-level engineering. It involves the development of earthquake-resistant drywalls. This discussion will help a reader recognize why there are subjects from diverse disciplines in the core engineering curriculum. For example, why should an electrical engineer undergo courses in chemistry, material science, and economics, or a metallurgical engineer learn about areas of mechanical and civil engineering?

The ideas will be developed in the following sequence: First, we discuss discipline-independent general thinking strategies and barriers to problem-solving with examples. Next, we discuss solutions to engineering problems, highlighting how these general thinking strategies and discipline-specific concepts are employed in the solution.

2 Problem Solving Strategies

Psychologists have identified the following strategies (Morgan et al., 1986) for solving problems independent of the discipline.

(1) Trial and Error
(2) Logical reasoning
(3) Divide and Conquer
(4) Representation / Modeling
(5) Analysis
(6) Reduction
(7) Reformulation
(8) Analogy

Analysis is a three-step approach to solving a problem: separating the whole into parts (divide and conquer), understanding the parts in isolation, and combining the understanding so obtained to understand the whole. The meaning of the other words above is self-evident. In this article, we will provide examples to highlight how each of the above-mentioned strategies can be adopted for arriving at solutions to different engineering problems.

Following two mental biases or barriers can prevent effective solutions to problems:

(1) mental set, and
(2) functional fixedness,

The mental set is a bias to stick with only the solution approaches that have worked in the past and stubbornly ignore alternatives. *Functional fixedness* is a bias that blocks people's ability to use an object or concept for something other than what it is usually used for (Morgan et al., 1986). One way of overcoming these mental biases is to become conscious of their presence, which is an exercise in self-awareness. Once an individual is aware of any limitations in their thinking, they may be able to avoid them through training their mind. One aspect of the training could be accepting that there can be more than one approach to solving a given problem. This acceptance brings about flexibility in thinking, eventually allowing the mind to become more open to possibilities. This also helps enhance the creative instincts of the individual. Another way of overcoming mental bias can be by developing a habit of solving different types of problems. This involves exposing oneself to different fields of work and improving the breadth of thinking. Good problem solvers are typically those individuals whose level of knowledge can be described by the letter "T", where the cap of the alphabet describes the breadth and the remaining part of the alphabet describes the depth of knowledge. This means that they have a general knowledge of many different topics but, at the same time, hold expertise or in-depth knowledge in one area or topic. Such an ability can be achieved by pursuing academic curriculums that provide a student sufficient exposure to topics of other disciplines besides his/her core discipline. Alternatively, one can take up different types of work assignments to develop their breadth of knowledge and, consequently, their problem-solving skills. These skills then become very useful for engineering problem-solving.

We shall discuss the above points with the help of four examples: solving an equation such as $e^{-x} = 0.5 + x^2$; tracing the reason why a computer is down; dividing a square piece of paper into five equal squares using scissors only; and discovering the structure of the atom.

2.1 Solution of the Equation $e^{-x} = 0.5 + x^2$

Experience shows that when presented with this problem, students instinctively look out for the use of computational software such as Mathematica, Matlab, etc. Indeed, they may even await the availability of such software. This is a *mental set*. They must

Table 1 Illustrating the solution of the equation $e^{-x} = 0.5 + x^2$ by trial and error and logical reasoning

x	LHS	RHS	Remark
0	1	0.5	LHS > RHS
1	0.368	1.5	LHS < RHS. Hence, x lies between 0 and 1
0.5	0.606	0.75	LHS < RHS. Hence, x lies between 0 and 0.5
0.25	0.779	0.562	LHS > RHS. Hence, x lies between 0.25 and 0.5
0.375	0.687	0.641	LHS > RHS. Hence, x lies between 0.375 and 0.5
0.438	0.645	0.691	LHS < RHS. Hence, x lies between 0.375 and 0.438
0.406	0.666	0.665	LHS$\approx$ RHS. Hence, 0.406 is the solution

realize that a simple calculator and the following *trial and error* approach guided suitably by *logical reasoning* can yield the solution in minutes. Let x_0 denote the solution of the above equation, i.e. the value of x for which LHS = RHS.

(1) Identify values x_1 and x_2 of x such that LHS > RHS for x_1 but LHS < RHS for x_2. We know that x_0 lies between x_1 and x_2. For the above equation, LHS rises as x falls; hence, LHS > RHS for x_1.
(2) Compare LHS and RHS for the average of x_1 and x_2, i.e. for $x = (x_1 + x_2)/2$. If LHS < RHS, x_0 lies between x_1 and $(x_1 + x_2)/2$, else x_0 lies between $(x_1 + x_2)/2$ and x_2. Suppose x_0 lies between x_1 and $(x_1 + x_2)/2$.
(3) Compare LHS and RHS for the average of these two values, and repeat the previous Step 2 until we arrive at an "average" for which the LHS differs from RHS by a predetermined small value. This average can then be regarded as the solution.

To apply the above procedure, we consider values $x_1 = 0$ and $x_2 = 1$, for which LHS and RHS are quickly evaluated. The solution can then be arrived at, as shown in Table 1. Thus, we have solved a fairly complex equation without the aid of any computer program, using a simple scientific calculator and a willingness to conduct a *logical trial-and-error* evaluation.

2.2 Why is the Computer Down?

Once, the desktop computer of a faculty member who belonged to a non-electrical sciences discipline stopped working. The faculty member regarded the computer as an electrical appliance and had a *mental set* (barrier) that discipline-specific knowledge of electrical engineering was required to resolve the problem. Hence, the faculty member requested that the author of this chapter, an electrical engineer, troubleshoot the problem. The author, however, could resolve the problem using discipline-independent general strategies in the following steps.

(1) The situation was r*epresented* using a diagram as in Fig. 1.

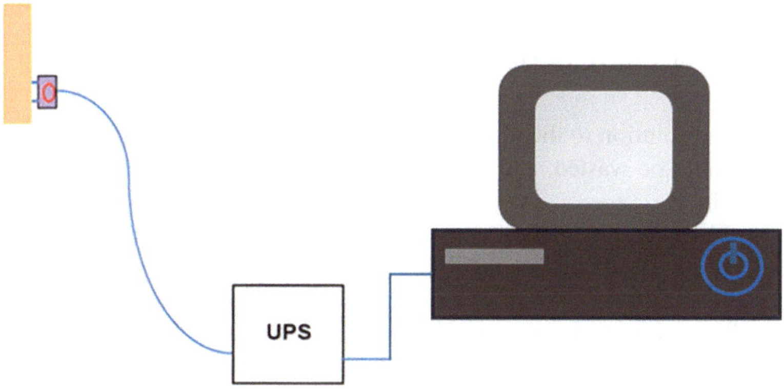

Fig. 1 A diagram describing the computer's connection to the power supply

(2) It was decided to *check the parts*, namely, the computer, wire segments, UPS, and plug and socket, *reasoning logically* that the computer could have been down because one or more of these parts had gone bad. This, in fact, meant *analy*zing the situation (refer to the meaning of analysis in Sect. 2).
(3) Visual inspection showed that not only the computer but also the UPS was down. This clearly meant a problem with the segment between the UPS and the socket. A problem might have developed in either the wire, the plug, or the socket.
(4) To check the socket, the plug was removed, and another appliance was plugged in. The appliance worked, clearly implying that the socket was sound and that the problem lay in either the plug or the wire.
(5) When the plug was opened for checking, it was found that one of the phases of the wire had become disconnected from the plug pin, perhaps due to frequent removal of the plug from the socket (there was only one socket in the room) to enable the connection of other appliances. Reconnecting the wire to the pin solved the problem, and the computer was working!

This example highlights the fact that to solve an engineering problem, before consulting a domain expert, one must attempt discipline-independent strategies, namely, *proper representation or model, logical reasoning, divide and conquer, analysis*, etc. An appropriate representation makes a problem more perceptible. Complex processes in industries are represented through process flow diagrams to facilitate troubleshooting (Hall, 2012).

2.3 Division of a Square Piece of Paper into Five Equal Squares Using Scissors Only

We consider a solution to the above problem assuming that cut and join is allowed, no paper should be wasted, and no paper can be pasted over another. Note that no scale is available.

The problem can be solved in several ways. First, we will describe an efficient way to solve the problem that highlights the role of the *reformulation* strategy in significantly reducing the solution effort. Next, we describe the three *mental sets* (barriers) that prevent or delay a solution to the problem. One of the authors of this article has presented the above problem to more than 10,000 students over two decades as an activity in a session on Critical Thinking as part of a course titled "Introduction to Research." It was noticed that most students fail to reach a solution because of these *mental sets*.

Figure 2(a) shows five equal small squares whose areas sum up to that of the large square. Please assume that the side of the small squares is 1 length unit; their area would be 1 area unit. The large square would have an area of 5 area units, implying its side would be $\sqrt{5}$ units long. We can *reformulate* the problem by "combining five equal squares into a large square" to reduce the problem *and* realize a length of $\sqrt{5}$ units starting from several 1-unit lengths. The following series of flashes with "aha" feelings lead to the solution.

(1) Pythagorean theorem learnt in secondary school tells us that $\sqrt{5}$ units is the hypotenuse of a right-angled triangle with sides 1 unit and 2 units long, i.e. $\sqrt{5} = \sqrt{(1^2 + 2^2)}$. (As aside, recent studies in the history of science and technology have shown that Baudhayana (800 BC—740 BC) of ancient India knew about this result a couple of hundred years before Pythagoras (570–495 BC) of Greece.) This *discipline-specific knowledge* of mathematics is required to solve the problem. In the author's experience, a couple of Humanities students of Aesthetics have also been able to solve the problem without resorting to this theorem from an intuitive idea of the shapes.

(2) See Fig. 2(b). We can join two small squares along their sides to get 1 unit and 2 unit lengths perpendicular to each other and a $\sqrt{5}$ units long "hypotenuse." A cut along this hypotenuse yields two $\sqrt{5}$-unit-long segments. Repeating this process with another pair of small squares yields two more $\sqrt{5}$-unit-long segments. Thus, we have realized four $\sqrt{5}$ units long sides of the large square.

(3) We rearrange the pieces to get the large square, yielding the four $\sqrt{5}$-unit-long segments as in Fig. 2(c).

In the author's experience, most of the over 10,000 students to whom this problem was presented failed to solve it in a reasonable time because of one or more of the following three *mental sets* (barriers) based on their long-term problem-solving habits.

- They take the problem statement as it is without attempting any *reformulation* or *reduction*,

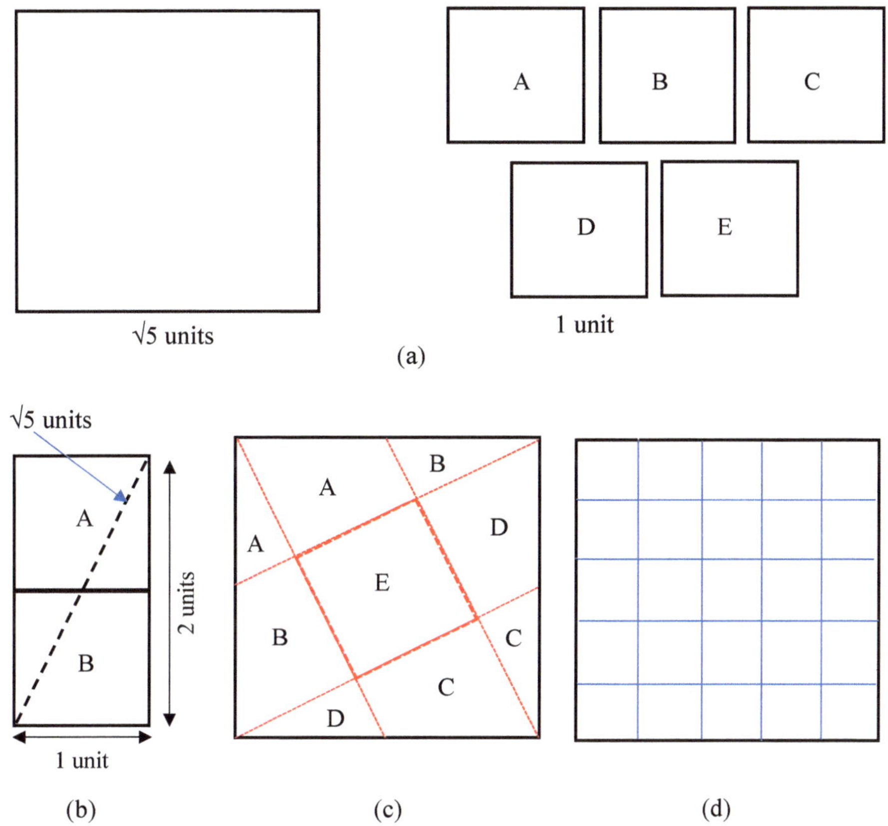

Fig. 2 **a** Representation of the problem. **b** Realization of an $\sqrt{5}$ unit long segment out of two small squares having 1 unit long sides. **c** Arranging the cut pieces of the small squares to make-up a large square. **d** Illustrating a *mental set* (barrier) of the division of the large square that cannot yield a solution

- They start dividing the paper along cutlines parallel to the sides. They vary the number of divisions of the paper but not the angle of the cutlines. Figure 2(d) shows a typical inappropriate division suggested.
- Among those who can *reformulate* and *reduce* the problem as discussed above, most conclude erroneously that a solution is not possible. This is because they believe that $\sqrt{5}$ is an irrational number, and such a length cannot be realized in practice. This is because they have always used a scale to measure length, on which $\sqrt{5}$ cannot be marked.

Note that even after we have *reduced* the problem and recognized the relative lengths of the large and small squares, as in Fig. 2(a), the solution can be delayed for a long time if we get *mentally set* to cut away a length of 1 unit out of $\sqrt{5}$ instead of *reformulating* the problem and trying to realize $\sqrt{5}$ units length from 1 unit length.

2.4 The Structure of an Atom

Research on the psychological processes underlying numerous discoveries, inventions and innovations has shown that most new knowledge is generated via *analogy*, i.e., transferring existing ideas to new surroundings (Brown & Selter, 2010, 167–169). An excellent example is the discovery of the atomic structure, where electrons revolve around the nucleus. This arrangement of the micro world was motivated by a prior observation of the planetary system in the macro world by the naked eye over 100's of years (Taber, 2013, 1377–1406). In the planetary system, planets revolve around a focal point, the Sun, in the case of our solar system. Scientists wondered why the essence of this arrangement of the macro world, where several entities revolve around a central entity, could not hold for the micro-world of atoms. Following up on this line, the scientists could explain the observations of the atomic phenomena by an arrangement of negatively charged particles called electrons revolving around a positively charged central entity called the nucleus. Thus, the Sun and planets of the planetary system are analogous to the nucleus and electrons of the atomic structure. Another example in physics is the discovery of matter waves from the wave-particle duality of electromagnetic waves.

3 Engineering Problems

Above, we discussed discipline-independent general thinking strategies and barriers to problem-solving with examples. Below, we discuss the solutions to two engineering problems, highlighting how these aspects and discipline-specific concepts are employed in the solution. One problem is developing a fabrication process for very high-frequency varactor tuning diodes with high-quality factors. Another problem is the development of earthquake-resistant drywalls.

3.1 Technology Development of Very High-Frequency Glass Packaged Varactor Diode with High Q

A diode is a two-terminal device that can rectify current or work as a varactor (variable capacitor). The capacitance varies when the DC voltage between the two terminals is varied. This device is used to vary the frequency during communication by radio waves up to 100 km in defence and other applications. As shown in Fig. 3, the diode consists of a semiconductor p–n junction chip of about 1 mm^2 area encapsulated in a glass tube sealed to two copper or alloy leads. A device with such an encapsulation can withstand extreme temperature and humidity.

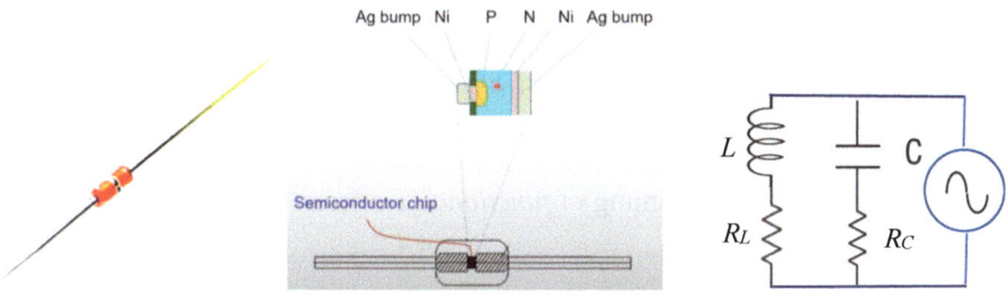

(a) (b) (c)

Fig. 3. **a** A diode. **b** Parts of the diode. The chip is held between two leads encapsulated in a glass tube. (c) The equivalent circuit of the Q measurement setup. C and R_C are the capacitance and parasitic resistance of the diode, while R_L is the parasitic resistance of the inductance L

environments encountered in defence applications. Here, we describe solutions to a couple of problems encountered in the technology development of a varactor diode.

3.1.1 Measurement of the Quality Factor (Q)

We start by reviewing *discipline-specific knowledge* relevant to this problem and its solution. The Q indicates the parasitic resistance, R_C, of the diode offered by the leads and the regions adjacent to the p–n junction. It is given by $Q = (2\pi f R_C C)^{-1}$, where f is the frequency at which R and the capacitance, C, of the diode, are measured. Thus, the higher the Q, the lower the R_C. The diode whose technology was developed has $C = 100$ pf and $Q > 400$ at 4 V reverse bias and $f = 50$ MHz. Although not pertinent to the specific problem discussed, we give the other details of the device for the sake of completeness: C is varied over 2 – 30 V and 30 – 300 MHz frequency range, device breakdown voltage > 30 V and a reverse current < 10 nA at 28 V reverse bias. To measure the Q of the diode, C is paralleled with an inductance L and an AC signal source, as in Fig. 3(c). If we use an inductor with resistance $< < R_C$, the current supplied by the source depends on R alone at the resonant frequency $f_0 = \frac{1}{2\pi\sqrt{LC}}$. We can estimate the Q at f_0 and a given reverse bias V_R from the measured current and C. In our case, $C = 100$ pF and $f_0 = 50$ MHz, so that $L \approx 100$ nH; $V_R = 4$ V.

During the development of the diode, the laboratory staff and faculty managing the equipment opined that the characterization of the device would have to wait for at least three months. This was because a readymade inductor of the required value and low resistance was not readily available and would have to be procured from the market. Overcoming this *mental set*, an author of this article pointed out that just a straight conducting wire of 0.1 mm diameter and 7 cm length will have a self-inductance of about 100 nH value so that it can be used (allaboutcircuits, n.d.). However, the series resistance of this wire was comparable to the diode R. Even this problem was overcome as follows. A device with a known high Q was measured using this straight wire as an inductor, and the Q measured in this manner was noted. An appropriate correction factor was derived and applied from these two Q values to get the accurate Q of the diode at hand from the Q measured using the straight wire inductor. In this manner, the delay in technology development was avoided.

Using a wire, generally used as a conductor to establish connections as an inductor, also demonstrates the overcoming of *functional fixedness.*

3.1.2 Silver Bump Electroplating

The relevant *discipline-specific knowledge* is as follows. Refer to Fig. 3(b). The leads press against the metal on the chip to make electrical contact. The pressure increases during the sealing of the glass tube to the leads at temperatures above 600 °C because of the contraction of the glass at this temperature. A good quality of contact results in this situation if the metal on the chip is soft and as thick as the diameter of a human hair. Silver deposited by electroplating achieves this objective. It is not practical to plate one chip at a time. Instead, hundreds of chips are plated simultaneously when they are part of a silicon wafer several inches in diameter before their separation by dicing the wafer. A schematic of the plating configuration is shown in Fig. 4(a). A voltage is applied between the silver anode and the silicon wafer. The resulting current causes plating.

After the setup was configured to plate the p-n junctions on the front of the wafer, it was observed that the wafer edges received much thicker plating than the desired p-regions of the junctions, as shown in Fig. 4(b). This was traced to the *mental set* (barrier) developed via numerous diagrams of the plating configuration from school days, which stated that electroplating is to be done using a DC voltage. A detailed (not just schematic) *model* or *representation* of the plating situation, as shown in Fig. 4 (c),(d), helped overcome the *mental set*. In Fig. 4(d), r_b and r_x are the resistances of

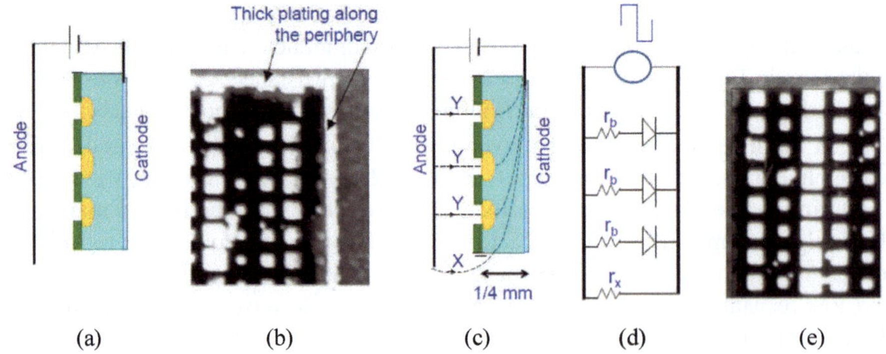

Fig. 4. **a** Wafer electroplating configuration. **b** Photograph of the wafer plated using DC voltage. **c** Current flow lines during plating. **d** An equivalent circuit is derived from (a) but with DC voltage replaced by AC voltage. **e** Photograph of the wafer plated using AC voltage

the paths Y and X in the plating bath shown in Fig. 4(c), and the diodes represent the p–n junction coming in series with the current path. The sum of r_b and the diode resistance is much higher than the resistance r_x of the path through the wafer edge. Since the voltage across all paths is the same, the current through the r_x is higher. This is why wafer edges receive more current and thicker plating than the p-regions.

The above thought process threw up an idea in the author's mind. If we use a square wave AC source instead of DC, the current through r_x will flow in opposite directions in the two halves of the AC voltage cycle, leading to a zero average current over one complete cycle. However, the diode allows the current in one direction alone. Hence, the average current through path r_b would be non-zero due to the rectifying nature of the diode. Figure 4(e) confirms that the use of AC plating voltage indeed prevents the plating along the edges while allowing the p-regions to plate as desired. This example highlights the role of accurate *representation or model* of the problem in overcoming a *mental set* and solving the problem. Interested readers can see (Karmalkar et al., 2002, 429–431), (Karmalkar & Sridhar, 2003) for more details of the above solution.

3.2 Earthquake-Resistant Plasterboard Systems

Partition walls in a house or apartment have traditionally been constructed with concrete. Concrete has been the preferred choice because of its high strength and durability. However, over time, it was realized that the high strength characteristic of concrete may be optional for non-load-bearing partition wall applications. As a result, real estate developers have gradually moved towards gypsum plasterboards to construct such partition walls. In addition to its lower carbon footprint, gypsum plasterboard demonstrates other advantages *vis-à-vis* concrete, such as ease of handling and significantly lower time to construction/installation.

Furthermore, in earthquake-prone regions, the gypsum plasterboard's lightweight nature reduces the building's damage due to seismic inertial forces. Material damage and human fatalities during an earthquake event are minimized if the various components of the building can survive the forces generated by the earthquake. In the case of partition walls, the displacement or strain to failure exhibited by the partition wall is considered a measure of its resistance to the seismic event. Generally, higher values of this metric are desired, and building codes such as ASCE 7 (asce.org, n.d.) have established displacement values that must be equalled or surpassed for a partition wall to be considered earthquake-resistant.

3.2.1 Mechanical Testing of the Plasterboard System

Now, we describe the *discipline-specific aspects* of the earthquake-resistant plasterboard system for partition wall application. The partition wall is constructed by first erecting skeletal metal (galvanized iron) components called channels and studs. Following this, as shown in Fig. 5, plasterboards are placed over and attached to the

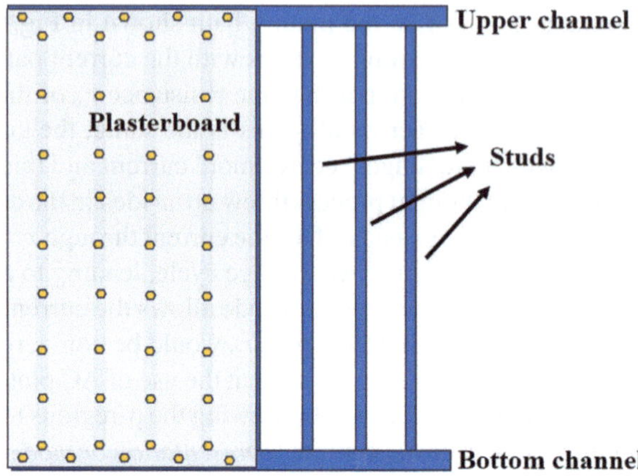

Fig. 5 A gypsum plasterboard-based drywall with a metal framework consisting of channels and studs for support. The plasterboard is attached to the metal framework through screws (shown by orange coloured dots in the above figure)

metal framework through screws. The displacement to failure of this plasterboard system (plasterboards and metal components) is dependent upon the strength and arrangement of the individual metal components, the type and number of plasterboards used in the construction of the drywall and the mechanical performance of the screws used in the attachment of the boards to the metal framework.

The company that one of the authors of this chapter was working for was interested in developing the next-generation earthquake-resistant plasterboard system. Several permutations and combinations of metal components, plasterboards and screws were thought of, and experimental evaluation of these possibilities was necessary for identifying the best earthquake-resistant plasterboard system.

The experimental campaign for evaluating the different combinations involved mechanical testing of full-scale partition walls. The Fig. 6 illustrates the full-scale wall used for testing. During testing, electro-mechanical actuators apply shear forces to this full-scale wall. Load and displacement sensors (LVDTs) are attached to the partition wall to capture the force–displacement response of the partition wall. The entire setup is quite elaborate, and a single experiment, from building the full-scale wall to its mechanical testing, could take as long as a week. Since time is of the essence in the commercialization of any product, the company was eager to complete the decided set of experiments in the shortest possible time.

However, these tests were being conducted at a third party's place, and hence, the long waiting times between tests, coupled with the significant time for each test, was delaying the product development process. It was then realized that the testing and evaluation process could be significantly accelerated if a *reduced* version of the partition wall was tested. A model of the partition wall, *analogous* to the full-scale wall, could be easily accommodated and tested on a universal testing machine. This machine is used to study materials' mechanical properties in different testing modes, such as uniaxial tension/compression and flexure.

Tests on the *reduced* model of the partition wall were then carried out on the universal testing machine. This approach paid immediate dividends as, within a few

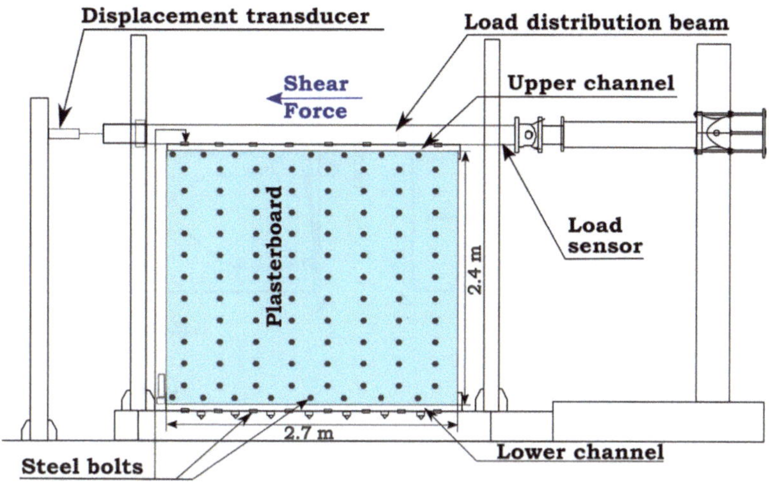

Fig. 6 Schematic showing the elaborate arrangement needed for conducting a shear test on a partition wall

weeks, all the different combinations of plasterboard systems had been evaluated on the *reduced* models. Interestingly, the results presented by these tests agreed with those obtained from full-scale walls. Figure 7 (a) and 7 (b) compare the force–displacement response obtained from the *reduced* model and full-scale wall tests. The trends indicate a good agreement between the two testing approaches (Gollapudi et al., 2020). This suggests that testing with a *reduced* model is a rapid and viable alternative to full-scale wall testing. Hence, the *reduced* model testing was employed to develop the earthquake-resistant plasterboard system.

The accelerated testing capability presented by the *reduced* model approach allowed improvements to the plasterboard system installation process. A new approach to installation was identified based on the introduction of an additional metal component called the metal insert. Integrating this metal insert with the metal framework provided test displacements superior to those specified by the building code ASCE 7 (Gollapudi et al., 2020). This can be noted in Fig. 7 (b), wherein the single-layer standard board with metal insert provided the highest displacement among the configurations tested. The company duly protected this new development through a patent filing.

The above example, therefore, highlights the effectiveness of the *reduction* strategy and *analogy* for solving engineering problems. By choosing a *reduced model* that was *analogous* to the full-scale wall, the testing process was simplified, allowing for faster evaluation of the planned tests. The above example also highlights the power of approximation as the *reduced* model testing, although an approximation of the full-scale test, facilitated the faster development of a product. In addition to these, the above example also describes the importance of staying flexible or open to new approaches to problem-solving. In the above example, the engineers

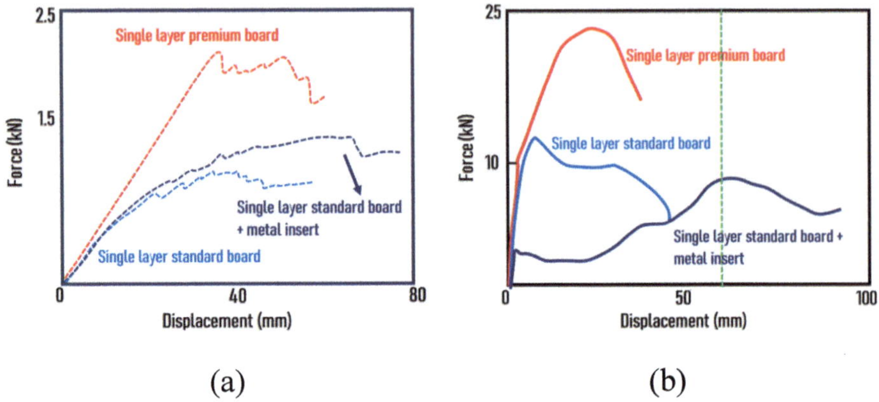

Fig. 7 Illustration of the force–displacement response presented by plasterboard systems from tests conducted on (**a**) *reduced* model and (**b**) full-scale walls. The trends obtained from full-scale tests are in good agreement with those obtained from *reduced* model tests. The green dotted line in Fig. 7(b) shows that with the metal insert in place, the performance of the plasterboard system exceeded the displacement specification laid down by the ASCE 7 code (asce.org, n.d)

could have been adamant against adopting the *reduced* model testing, stating that the data generated from the small-scale tests were not scientific enough. However, the engineers could accelerate the product development process by not having a mental set against the reduced model test. It is also important to mention herein that this product's successful development resulted from a collaborative exercise carried out by engineers from metallurgical, civil, and mechanical engineering. This reiterates our belief that effective engineering problem-solving requires an interdisciplinary approach, and people of different backgrounds must be encouraged to come and work together.

4 Conclusion

In this chapter, we discuss different strategies for solving engineering problems. These strategies were elucidated by discussing their applicability to both general and discipline-specific engineering problems. These strategies are crucial to engineering problem solving: *trial and error*, *logical reasoning*, *divide and conquer*, *representation/modelling*, *reformulation*, *analysis*, *analogy*, and *reduction*. After demonstrating the success of these strategies in solving general problems, such as determining the solution to a complicated equation or cutting a square into five equal squares, their role in solving real engineering problems was discussed through examples of very high-frequency glass packaged varactor diode with high-quality factor and earthquake resistant plasterboard systems. Indeed, complicated engineering problems can be resolved by effectively using these strategies. Furthermore, it is essential to be aware of the challenges to problem-solving brought about by *mental set* or *functional fixedness*. These biases or functional fixedness have to be overcome for successful engineering problem-solving.

References

Brown, S., & Selter, S. (2010). Analogies in science and science teaching. *Advances in Physiology Education, 34*, 167–169.

Gollapudi, S., & Kalanad, A., Bulusu, S., Sharma, M. (2020). *Earthquake-resistant construction element*, US Patent, 1066, 973, 3B2.

Hall, S. (2012). *Rules of thumb for chemical engineers*. Butterworth-Heinemann.

Karmalkar, S., Venkatsubramaniam, N., & Oak, S. (2002). DC and periodic reverse electroplating of semiconductor surfaces having adjacent P-type and N-type areas. *Journal of Electrochem. Society, 149*, C429–C431.

Karmalkar, S., & Sridhar, D. (2003, August 1). A method of electroplating metal contacts on a P+N semiconductor diode substrate and an apparatus for carrying out the said method. Indian Patent # 188350 (sealed on).

Morgan, C., King, R. A., Weisz, J. R., & Schopler, J. (1986). *Introduction to psychology*. McGraw Hill International Edition.

Taber, K. S. (2013). Upper secondary students' understanding of the basic physical interactions in analogous atomic and solar systems. *Research in Science Education, 43*, 1377–1406.

https://www.allaboutcircuits.com/tools/wire-self-inductance-calculator/.n.d. Accessed April 6, 2024.

https://www.asce.org/publications-and-news/asce-7. Minimum design loads and associated criteria for buildings and other structures (ASCE / SEI 7–22). n.d. Accessed April 6, 2024.

Chapter 5
Immersive STEM Education

Adrian McCallum, Vikki Schaffer, and Brendon Munge

Abstract What key attributes make a graduate career-ready? This chapter delves into immersive learning as a critical methodology for supporting career readiness. Immersive learning is characterized by direct workplace, laboratory, or field engagement. It is an effective approach to cultivate these attributes. Through practical examples, we illustrate how immersive experiences can bridge the gap between theoretical knowledge and practical application, thereby enhancing the readiness of graduates for professional challenges. Examining recent practices and educational doctrines to envisage the future of STEM education, the focus is on reimagining STEM learning pathways to better align with industry needs and support graduates' adaptation to the rapidly changing professional landscape. Significant shifts in educational strategies are essential to achieve this goal. These shifts should prioritize experiential learning, adaptability, and deeper industry-academia collaboration. Concluding, we posit that to meet the dynamic demands of the modern workforce, a re-evaluation of current educational practices is imperative. We advocate for large-scale changes in preparing graduates, emphasizing the importance of immersive learning experiences and a forward-thinking approach to STEM education. This will ensure that future graduates are well-equipped to navigate and thrive in an ever-evolving industry environment.

1 Introduction

Learning is best accomplished through physical immersion in the desired environment. Immersion can take many forms, including use of technology, such as virtual and augmented reality headsets (Markowitz et al., 2018). However, outside the virtual world, the concept of reality is critical and realistic (like real) experiences, can be coordinated in the classroom and beyond. Immersion is described as complete involvement and deep engagement in realistic experiences, enhancing training and deeper conceptual thinking (Coomans & Timmermans, 1997; de Freitas & Neumann,

A. McCallum (✉) · V. Schaffer · B. Munge
The University of the Sunshine Coast, Sippy Downs, QLD, Australia
e-mail: amccallu@usc.edu.au

I. Gratchev et al. (eds.), *Transformative Approaches to Career-Ready Education*,
https://doi.org/10.1007/978-981-96-7018-5_5

2009). This process, crucial for learning, enables more profound engagement with information, fostering enhanced mental capabilities and stimulating absorption in activities and studies (Feinstein et al., 2002). Immersive experiences in real-world or active learning settings are vital for facilitating engagement and learning enhancement.

In this chapter, we briefly examine the concepts of immersive learning in STEM education, the role that field or *in-situ* immersion plays in this process and how emerging technology may enable increased diverse and inclusive involvement when field visits are not possible.

The benefits of 'active learning' or 'learning by doing' will be discussed in the context of immersive engineering laboratory exercises and project management experiences by students undertaking regular site visits and engagement with domestic and international industry stakeholders as part of ab initio STEM undergraduate programs. In examining and discussing the numerous beneficial ways in which real immersive pedagogical techniques might be utilized within undergraduate and post-graduate STEM tertiary education, we will highlight the benefits of these approaches in developing 'beyond-classroom' skills that equip graduates with career-ready skills far more significant than they might otherwise attain, through typical classroom learning.

Experiential learning can be described as "any component of the curriculum that involves leaving the classroom and learning through first-hand" or immersive experiences (Boyle et al., 2007, p. 300). Such learning is best accomplished through physical immersion in the desired environment. Munge et al. (2018) reviewed the literature on experiential learning, specifically outdoor fieldwork in higher education. They found that immersive learning in outdoor fieldwork contributes to achieving affective, cognitive, and skills-based learning, contributing to student's overall professional development, identity, engagement and retention in their studies and future careers.

The challenges of immersive learning, such as time intensity, cost, accommodation, and logistics, are amplified in remote locations. Physical dangers such as wildlife, extreme weather conditions, or anthropocentric risks (political instability and conflict zones) must be addressed (Munge et al., 2018). However, 'fieldwork' (in all its forms) is vital for gathering rich, context-specific data and experiences that are hard to obtain through other means.

In addition to examining and discussing the numerous beneficial ways immersive pedagogical techniques might be utilized within undergraduate and post-graduate tertiary education, we confirm the notion of a 'career-ready graduate' by examining the concepts and benefits of immersive learning techniques. Furthermore, we look briefly at where immersive teaching might be heading, both within the university context and sectorally, by examining the continued role (or not) that universities might play in preparing career-ready graduates.

Before examining the concepts, benefits, and future of immersive STEM education, it is valuable to briefly revisit the rationale of such education: What do employers expect from STEM-qualified career-ready graduates?

2 What Is a Career-Ready Graduate?

Testa (2024), in a blog for Labsta, noted that career readiness was gaining prominence globally, with employers wanting empowered graduates with the skills, knowledge and attitudes required to succeed in their chosen careers. These attributes include foundational (and perhaps typical) STEM skills in reading, writing and critical thinking, technical skills such as technical specialization and programming literacy, and more nuanced skills known as "soft skills" or "transferable skills". These skills are critical to gaining employment and career longevity as soft skills are transferable, remain with us and can be drawn upon in any situation. Testa also highlighted the importance of self-reflection and the desire for work-based learning experiences to enhance an individual's ability to apply theoretical knowledge practically in a real-world environment. Karimi and Pina (2021) highlight that research has continuously demonstrated the importance of transferable skills for several decades. The broad nature of soft or transferrable skills from teamwork to reflective practice means considerable overlap exists. Soft skills can be characterized as interpersonal, such as listening, leading, teamwork and conflict management or intrapersonal, which include reflective practice, perseverance, task, time and self-management (Colburn, 2018).

In an Australian study, Prikshat et al. (2020) identified that many stakeholders perceived that graduates from Australian universities needed to be work-ready and needed more of the skills identified by Testa (2024). A lack of authentic learning opportunities characterized by the need for work-ready graduates is linked to a diminishing requirement for work-integrated learning programs and a reduction in stakeholders partnering with HE institutions to offer work placements (Prikshat et al., 2020). To reduce these deficiencies, the authors proposed that universities review their pedagogical focus on experiential and interactive immersive learning. The importance of incorporating avenues for students to develop soft skills is evident based on growing demand from employers (Sheedy, 2023).

Caten et al. (2019) reiterated that soft skills such as leadership, creativity, communication, management, professionalism, ethics, agility, resilience, and flexibility in Engineers, for example, exceed the importance of technical skills.

Engineers Australia lists five essential soft skills that graduates should acquire: time management, project management, how to write an email, grammar, and communications skills. These skills are not always evident at the forefront of STEM courses/subjects or program/degree outlines. We suggest a shift away from the concept of hard and soft skills due to the perception it can create related to the difficulty of the skills. Instead, utilize the language of affective, cognitive, and technical skills to encompass a graduate's characteristics (Munge et al., 2018).

Graduates with skills that encompass affective, cognitive, and technical skills can bring new insights while continuing to learn. In-depth knowledge and understanding of principles, theories, and practical applications, combined with the ability to collect, analyze, and interpret data, are essential for making informed decisions. The ability to leverage technology to solve complex problems, think clearly and rationally, understand the logical connection between ideas, solve problems innovatively, be open to change, and adapt to demands can also be developed. Other desirable career-ready skills include effective communication, the ability to explain complex concepts to non-specialists and the ability to collaborate effectively with teammates. These skills include sustainability literacy, cultural understanding, and working effectively in multidisciplinary, multi-cultural teams to achieve common goals; additionally, the capacity to think outside the box, to identify and frame problems, and to generate innovative solutions is crucial.

The range of attributes that industry seeks from STEM graduates may be wide and varied. However, as access to content and technical knowledge becomes more immediate, employers typically seek 'background' affective and cognitive skills more readily. Having briefly examined what a career-ready graduate looks like, we now examine some concepts of immersive/hands-on learning that may be useful in preparing such a graduate.

3 Concepts of Immersive Learning

In this section, we examine concepts of immersive learning, using examples from the authors' experiences, that provide some foundations for our immersive teaching.

Learning by doing is not a new concept. McCallum's indoctrination into 'active learning' or learning by 'doing' started over thirty years ago when he joined the Australian Defence Forces as an Officer Cadet. Amidst university studies, many other military subjects were examined, highlighting the impact and effectiveness of this pedagogical method. You don't learn how to re-assemble a Self-loading Rifle (SLR) blindfolded in less than 60 s by listening to a lecture or reading a book; you learn it by 'doing'. Children learn by watching and then imitating their parents or by observational learning (Bandura, 2004); learning by doing is the basis of our evolution as humans, yet why is it sometimes ignored in contemporary tertiary education (Mills & Treagust, 2003)?

In addition to necessary military training, McCallum's formative military days also introduced him to the concept of Adventurous Training, a technique designed to develop skills and encourage resilience that draws on the theory of adventure-based learning (Australian Army, 2018); it is a sometimes-extreme example of 'reality-based experiential learning' (Barraclough, 2016). Again, this immersive (in this case, as close to war as possible) learning stimulates attitudes and behaviours that mirror those desired in times of conflict: rapid decision-making amidst poor communication, with limited information, teamwork, and assertiveness.

In a civilian context, contemporary application of these techniques is practised by organizations such as Outward-Bound Australia (OBA) (McKenzie, 2003) and McCallum recently opined on the inherent alignment of his pedagogical philosophies with those of OBA via an online article (McCallum, 2020). Exploration of

McCallum's application of these methods will be examined shortly, but they align with the general benefits of outdoor fieldwork (Munge et al., 2018).

Through field-based experiential learning, students engage in activities that push them out of their comfort zones, challenging them to solve real-world problems, work in teams, and navigate complex situations. These experiences help students better understand their strengths, weaknesses, interests, and values, fostering greater self-awareness and personal growth. Fieldwork, as a form of experiential learning, entails curriculum components that require students to learn outside the classroom through immersive experiences (Munge & Thomas, 2021). Recognized for fostering deep learning (Munge et al., 2018; Hill & Woodland, 2002), fieldwork enhances student engagement and connects students with their areas of study. Fieldwork encourages personal growth, responsibility, tolerance, and social inclusion. It allows for exploring new environments and perspectives, contributing to affective learning by fostering a sense of wonder. Fieldwork offers new perspectives of the world that enable personal development and self-confidence, a sense of responsibility and tolerance, and opportunities to explore and investigate (Field Studies Council, 2012). While opinions on the effectiveness of fieldwork in enhancing student skills vary, it is widely acknowledged for supporting the development of general and subject-specific skills, solidifying theoretical knowledge, fostering intuitive understanding, and developing tacit or intuitive knowledge (Diaz et al., 2015; Jackson, 2013; Ribeiro, 2012). Fieldwork facilitates personal development, boosts self-confidence, helps overcome fears, and promotes social integration.

For many disciplines, fieldwork becomes their signature pedagogy (Shulman, 2005; Thomas, 2015). The Geoscience disciplines, Outdoor and Environmental Education, and Forestry utilize immersive fieldwork as a pedagogical framework to synthesize and teach the essential components of the theory, knowledge, skills, and practical applications underpinning their discipline's professional characteristics. The fieldwork allows for authentic experiences that replicate the professional context of the field. Importantly, pedagogical frameworks exist via support by experienced staff that aids and supports students as they grapple with the synthesis of theory and practice (Munge et al., 2018). Fieldwork can be complex and 'unwieldy or messy' to accomplish, which brings about a key component of authenticity linked to real-world applications; however, educators can mitigate many of these matters with thorough preparation, ethical considerations, and cultural sensitivity (Munge & Thomas, 2021).

4 A Step into Immersive Course Design

Stepping back from the field, learning by doing is successfully practised on campus by many. John Atkinson, Emeritus Professor of Soil Mechanics at City University, London, promotes the pedagogical philosophy of "learning by doing" (2012, 2016), and in Australia, similar practice within engineering education is also advocated by Jaksa et al. (2012).

McCallum et al. (2016) documented the conduct and benefits of immersive classes in Engineering at the University of the Sunshine Coast through the development of courses such as Soil Mechanics and Advanced Soil Mechanics.

Advanced Soil Mechanics was established as a project-based course that 'employs' the students as 'Project Engineers' to plan, design, construct and test an engineering structure. Work is rigorously assessed via group and individual project reports and briefings. These capture how students planned and conducted their practical soil investigations, what data they found, and what design and engineering implications arise from such data. In groups or teams, students verify field-collected data with additional laboratory testing before applying their findings to design and test this 'real' engineering structure under typical and 'worst-case' design loads. This course was designed to encourage students to take responsibility for their learning (a final/4th-year course) to ease their subsequent transition into the workforce, where they will be expected to carry out similar tasks independently.

Career-ready facets of the course are that it is:

1. Learning-centred: Learning by 'doing' focuses on supporting and enabling learning for all students, particularly by encouraging active and collaborative learning, where students also teach one another (Vygotsky & Cole, 1978).
2. Standards-based: Learning based on 'doing' can be founded on Engineers Australia competencies such as the application of established engineering methods to complex engineering problem solving (Engineers Australia, 2017); required professional requirements such as effective oral and written communication, and effective working in teams (Engineers Australia, 2017), can be embedded via group problem solving and reporting activities, and most importantly it is,
3. Career and future-focused: Enhancing (dynamic) skills rather than (passive) content learning by 'doing' ensures that graduates are career and future-focused.

However, emphasizing learning by doing and adopting career-relevant skills should not be left to final, end-of-degree courses. These skills and inclinations should be introduced as early as possible. As theorized by philosopher, psychologist, and educational reformer Dewey (1938), learning by doing is a pedagogical method based on the premise that learning can be enhanced.

An additional workplace-immersive course (not strictly WIL), Foundations of Engineering (ENG101; now Professional Engineering), was developed to enhance student retention and engagement within Engineering. Workshops with staff and students were conducted to determine the desired course format and content, and numerous criteria were identified that should be incorporated to create an immersive learning environment that encouraged students to be career-focused from the commencement of their Engineering studies. Criteria included:

- Provide an engaging hands-on experience in the spirit of the conceive, design, implement, operate (CDIO) model or similar (Crawley et al., 2007).
- Explicitly establish the principles and framework for engineering design and problem-solving.

- Explicitly establish the principles and methods of practical group work through activities that assist cohort building.
- Follow a structured and activity-based format that motivates students to attend because they are doing fun things and cannot simply watch the lecture and skip the class.
- Avoid large weighted assessment tasks and instead follow a modular approach with some marks given to small tasks linked to the workshops and lectures (for instance, the tasks might be linked to a portfolio of drawings and problem-solving tasks).
- Avoid the use of exams, although small quizzes may be appropriate, and
- Consider previous student feedback on the course and involve students as partners in the redevelopment.

These desired components largely align with the concept of 'active learning', and after the course was redesigned to incorporate these components, student feedback was effusive. This type of learning aligns with Kolb's (2015) four-stage cycle of experiential learning whereby the integration of four learning modes precipitates learning: (i) experiencing, (ii) reflecting, (iii) thinking and (iv) acting, while also considering the different student learning styles.

5 Beyond Engineering

At one Australian university, Anatomy classes have been revolutionized to create an immersive learning environment (Shaping Australia, excerpt from The Australian newspaper, 28 February 2024). Teaching revolves around practical workshop activities involving continuous student interaction. Plasticized specimens, body specimens, and human bodies enable hands-on learning and provide exits for uncomfortable students to watch live-streamed examinations from an adjoining room. This innovation has resulted in highly positive student feedback from students and health professionals.

Work-integrated learning (WIL) is another popular form of tertiary learning that specifically encapsulates learning techniques designed to generate career-ready graduates. In Australia, Universities Australia's report, Career Ready Graduates (2019), highlighted WIL benefits, defining WIL activities as work placements and internships, industry projects, fieldwork, and industry simulations. The benefits of WIL can be substantial and are examined in the next section.

Immersive learning may be described and implemented in various ways, many of which are similar (but not necessarily the same): active learning (espoused here), CDIO, PBL (project-based learning), and WIL. Next, we briefly examine some additional benefits of these approaches.

6 Benefits of Immersive Learning

Many studies document the benefits of immersive learning, both within and outside a university context. Schaffer (2017) reiterated that effective teaching and learning should offer equitable and authentic experiences that meet diverse student needs and goals, primarily through student engagement. Authentic learning and assessment and experiential learning—which connects directly with real-world problems—support deep cognitive skills and meaningful learning that integrates key concepts, previous and new knowledge, and personal experiences (Hunt, 2012). Learning environments extending beyond the classroom cater to varied learning styles and foster real-world connections for a deeper understanding (Kolb & Kolb, 2005). Experiential learning bridges the gap between academic knowledge and practical application (Ruhanen, 2006).

Additionally, social constructivism emphasizes students' individuality and active involvement in learning, supporting mixed-modal approaches that blend physical and virtual experiences for immersive engagement. Immersion, characterized by complete involvement and deep engagement in realistic experiences, enhances learning by providing real-life contexts for training or conceptual thinking, promoting deeper mental processing and absorption in the learning material (Evans & Acton, 2022).

These skills can be enhanced through work-integrated learning (WIL).

Many authors, such as Mandal and Edwards (2022), have highlighted the value WIL can bring to provide students with work-ready skills. WIL is commonly practised and is a particularly beneficial form of immersive learning. It is critical for developing broader skills such as individual and career self-management, networking, and applicable skill development of discipline knowledge (Jackson, 2018). WIL is another form of tertiary learning that specifically encapsulates learning techniques designed to generate career-ready graduates. A University of Australia's Career Ready Graduates (2019) report highlighted that WIL activities could include work placements and internships, industry projects, fieldwork (locally, nationally, or internationally), and industry simulations. This further relates to work readiness, "the ability to function effectively upon entering the workforce and across a range of contemporary working environments" (Jackson, 2018). Career or work readiness incorporates the diverse capabilities and attributes graduates need to successfully enter job markets that are dynamic, evolving and, at times, precarious work environments (Campbell & Price, 2016).

Another beneficial immersive learning strategy is deep learning. Deep learning involves actively engaging with the material, leading to insights and connections, contrasting with a superficial approach that depends on memorization. In gaming, Jennett et al. (2008) describe immersion as a multi-level experience, starting with engagement (investing time, effort, and attention), progressing to engrossment (emotional involvement), and culminating in total immersion, characterized by presence in a virtual environment and high involvement. Achieving total immersion in students necessitates appealing and relevant teaching methods.

Co-presence in learning and socialization is another appealing learning method. According to Haywood and Cairns (2005), socialization plays a crucial role in the learning process, although it may involve something other than active cooperation or collaboration. The essence of co-presence focuses on mutual awareness among participants that may not require interacting directly. The presence or feeling of being with others, even without direct interaction, enhances the psychological experience of being in a particular space. This enhanced presence can lead to greater opportunities for engagement, as Torisu (2016) noted, suggesting that the psychological component of feeling connected or being in the company of others can significantly influence levels of engagement with the environment or activity at hand.

Moving back to the benefits that field or outdoor experiences can bring, Kirk (2021) broadly articulates the life-changing benefits that adventure can bring, highlighting facets like independence and resilience that may help shape a more effective and career-ready university graduate. Munge et al. (2018) found that outdoor fieldwork experiences provided rich, immersive experiences that encapsulated student engagement in authentic work-orientated tasks that built professional competencies while establishing socialization and enculturation into workplace expectations.

McCallum (2024; unpublished) recently returned from two overseas immersive learning experiences with undergraduate engineering students. Student reflections overwhelmingly supported the value of such immersive 'living-it' education, where real-world problems were encountered and examined in real-time, providing students with exposure to a far greater range of skills, such as communications, expectation management, and stakeholder management, that are evident in any real-world engineering project but are never encountered during a typical undergraduate engineering degree, except perhaps via (often supported) management of a 4th year engineering research project. Similar perspectives were also identified in an earlier similarly immersive education program in Nepal (McCallum, 2017; unpublished). When students travel and are immersed in new places and circumstances, learning can occur serendipitously with new concepts, opinions and skills derived from exposure to other people, cultures, disciplinary knowledge, and industry practices (Berg et al., 2023). Berg et al. (2023) highlight the importance of accompanying educators to support directed student learning within such immersive educational experiences. Emergent learning also falls under the learning model of experiential learning.

Experiential learning, or learning by experience, encourages students to draw on their previous and current life events and those revealed from participating in educational-based activities (Andresen et al., 2020). The key to gaining benefits from experiential learning is analyzing the learner's experience through self-reflection, evaluation, and reconstruction to extract meaning (Andresen et al., 2020). Self-reflection is another often neglected 'soft skill' that should increasingly be used to imbue in students a cycle of self-examination and improvement that will assist in their self-awareness and immediate capability within the workforce (McCallum & Fairweather, 2019).

Australian tertiary, multidisciplinary field experiences (2015 and 2017) in the Pacific Islands of Vavau, Kingdom of Tonga, included a daily debrief where students reflected on the day's activities, including engaging with the community. By drawing on the daily and overall field activities, students drew on their existing understanding to further understand and increase their knowledge of their disciplinary fields and those of their peers. By evaluating their experiences, they can stimulate critical thinking, which may result in attitudinal change and intercultural understanding (Andresen et al., 2020).

Professional bodies such as Engineers Australia increasingly call for evidence of self-reflection within undergraduate curricula, and we (as well as others Hermsen et al., 2022) argue that establishing the practice of self-reflection (in itself an active, reflective process, rather than a passive content-receiving process) in all types of education, assists in the development of graduates who are better placed to examine their professional workplace efficacy, and assertively initiate training or modifications to ensure ongoing professional effectiveness.

Having examined some of the benefits that immersive learning experiences can bring, we now briefly examine emerging doctrine and pedagogy and postulate on other immersive learning techniques that could be used to prepare STEM students for careers successfully.

7 Future Immersive Learning?

In the next section, we briefly examine personal and broader perspectives about how learning can better prepare individuals to understand themselves and their motivations and thus better select and apply themselves in a career after tertiary graduation.

McCallum (2020), after substantial self-reflection on the 'purpose of life', resolved to develop an immersive tertiary education learning experience that could be applied globally by any tertiary education provider within any (appropriate) external field environment.

His conception was: "Empowering journeys of education and discovery for the mind, body and soul" that draw upon three things:

1. Intensive tertiary-education offerings,
2. Adventurous travel, as offered by many adventure travel companies and
3. Immersive opportunities for self-discovery, as enabled by Outward Bound etc.

By combining these three facets within an intensive, expedition-style tertiary education experience, the following outcomes could be delivered:

- Tertiary Education Quality Standards—Australia (TEQSA) recognized intensive tertiary education (of a particular nature, about the expedition environment: desert, rainforest, coastal, alpine, polar, etc.),
- A physically challenging, immersive expedition experience that will promote team building, communication, and camaraderie (soft skills), which will also,
- Stimulate and enable opportunities for self-reflection, self-discovery and empowerment.

Students would be supported to push themselves physically while growing in self-confidence and obtaining a tertiary qualification, thus growing in "mind, body, and soul."

McCallum undertook a process of contemporary peer engagement to realize this initiative of teaching, learning, and growth that extended beyond institutional and national boundaries because of the international application he envisaged.

After international consultation and further reflection, McCallum wrote a self-reflective article that enabled him to synthesize his reflections and his conversations in devising this teaching and learning initiative.

While writing this piece, he realized that the organization Outward-Bound primarily addressed what he sought to achieve through this 'social initiative' (McKenzie, 2003), and he conferred with their Board and CEO on how to replicate their experiences in a tertiary education environment.

Thus, after having conferred at length with colleagues at all levels, both within and beyond his current institution, McCallum delivered an encapsulation of his original concept, a revolutionary teaching and learning initiative that not only provides students with tertiary-level education but also provides for them an immersive, active-learning experience, founded upon proven concepts (Australian Army, 2018; McKenzie, 2003), through which they may experience personal revelation and growth that might empower them, beyond their tertiary education experiences.

Beyond McCallum's contemporary personal reflective insights, many other evolving factors should be considered when considering future immersive learning.

Additional tools to ensure the graduation of work-ready graduates could be enhanced by industry/academia exchange opportunities. Drawing upon field experience and work-integrated learning models, industry/academia exchange could enable experiential learning for industry professionals who contribute to academic courses/subjects for a semester and academics who temporarily fill the industry role (part-time or full-time). Both participants can 'up-skill' to attain contemporary industry experience that can be incorporated into teaching. These teaching-to-professional exchanges can be supported by tools such as embedded industry PhDs (or similar) so that future academic staff a. learn and progress their research within an industry environment, whilst b. they are developing embedded networks and relationships with industry, ensuring that ongoing collaboration is more accessible, reducing often present silos and isolation.

The future-readiness of graduates is also often touted. Higher education aims to incorporate employability skill development within courses and programs to assist students to ready themselves "for jobs that do not exist yet, using technologies that have yet to be invented, and for solving problems that nobody has yet thought of" (Kumar 2007 cited in Römgens et al., 2020, p. 2588). Römgens et al. (2020, p. 2590) suggest that at its core, employability refers to "an individual's perceived ability to obtain and maintain employment throughout his/her career". These authors also reveal that employability within tertiary education is evaluated on individual competencies based on the knowledge, skills and attitudes that contribute to effective work-based performance. These competencies vary by institution and discipline and, therefore, can be challenging as a tool/approach to understanding what has been learned or achieved in this context. Innovative approaches within higher education require reimagining theoretical and practical avenues to help students navigate the uncertainties, opportunities, and challenges ahead.

Educational gaming platforms like Kahoot!, Minecraft: Education Edition and Prodigy use game mechanics to make learning fun and engaging. These platforms cover various subjects and skills, offering interactive challenges that promote critical thinking and problem-solving. Mixed modality can link virtual or remote field trips to offer meaningful collaboration between on-campus students, in-field students, and other stakeholders; VR and 360° video technologies can provide immersive experiences of distant or inaccessible locations. Tools like Google Cardboard, Oculus headsets, and 360° cameras enable students to explore virtual environments in situ or virtually to create/recreate physical presence. Integrating interactive or non-interactive technologies for immersion significantly enhances learning in fieldwork by making data collection more efficient, engaging students with interactive tools, and providing new perspectives on the subject matter (Zhao et al., 2023).

Any discussion on reimagining education would be complete with contemplating changes to the sector due to AI. Rapid technological advancements have left many institutions playing 'catch-up'. Educational institutions must be highly cognizant of stakeholders' changing learning expectations and outcomes, including students, industry, government, and communities. Course redesign and development increasingly incorporate digital skills. Akour and Alenezi (2022, p. 784) suggest that the emphasis on digitalization reduces the application of more "integrated, collaborative, multidisciplinary, and cross-cultural learning approaches". In addition to educational institutions, workplaces have also been, and continue to be, changed by digital transformation. This has increased attention on emotional intelligence, with research exploring non-cognitive intrapersonal and interpersonal skills within educational learning objectives and the potential impact this may have on professional efficiency and achieving intended outcomes (Cunningham & Villaseñor, 2016).

For example, Jones-Vlasceanu (2023) examined Artificial Intelligence's (AI) transformation of career preparation and noted that students "will need to learn the skills of new career readiness, especially self-knowledge and self-efficacy" and that these skills can best be learnt by offering students "skills development, professional development, and experiential learning." Encouragingly, although the skills required in an AI-dominated work environment will change, how these skills can best be attained aligns with our thesis herein. Continuing advancements in AI, as well as Information Technology, Augmented Reality (AR) and Virtual Reality (VR), and Robotic Automation, are transforming fields like Engineering that require career-ready graduates with the technological and digital skills to support evolving and dynamic economies and imperative (Bogosian et al., 2020).

Similarly, serious games could be incorporated into education to support learning and develop the skills required for advanced digital literacy. Serious games utilize machine learning, such as image and voice/speech recognition (Mostafa et al., 2020) and also draw on emotion to personalize educational games to facilitate real-time responses (Westera et al., 2020). Game-based, immersive technologies can enhance teaching and learning in various STEM disciplines, such as mathematics, engineering, physics, chemistry, biology, and astronomy (Dyulicheva & Glazieva, 2022). Supplementing fieldwork, tools such as Learner-Immersed Virtual Interactive Expedition (LIVIE) leverage SV-IVR to integrate immersive and interactive virtual inquiry-based fieldwork into the learning and teaching of physical geography. However, the psychological effect of technology-supported learning suggests these are supplementary tools that, considering the social nature of learning, do not entirely replace human contact (Bansal et al., 2022; Verner et al., 2022).

In this section, we have discussed some possible options for future immersive learning in universities, but how guaranteed is such a future role for these institutions?

8 Where Is Higher Education Heading?

Higher education is rapidly evolving, requiring a re-evaluation and new approaches that facilitate autonomy and agency in student learning, the utilization of mixed contexts (de Freitas & Neumann, 2009) such as virtual reality (VR), recreated and replicated reality and physical immersion opens avenues for experientially authentic learning experiences (Lauria & Morie, 2015) that allow students to develop graduate attributes and career ready skills.

However, industry, which requires career-ready graduates, suggests by its actions that universities may need to move more quickly to address these requirements.

In 2018, McCallum (2018), reflecting upon Davis' work (Davis, 2017), observed that in the absence of universities producing career-ready graduates, private tertiary education providers were stepping into the void to ensure aligned training of future employees in a controlled and timely manner.

Further, some parties (personal communication McCallum, 1 March 2024) suggest that university education could be abolished entirely for many historically vocational courses such as teaching. Instead of spending three or four years at university gaining a teaching degree, prospective teachers could spend two to three years at a specific teachers' college, where they might study the art and craft of teaching for, say, three days per week whilst spending the remaining two days of their working week within a school, learning on the job.

Career-ready teachers and STEM professionals might learn career skills from their first week of education. Such an examination across the whole range of STEM education could be valuable, aligned with the current notion of whether university is necessary for future professionals; the World Economic Forum (2020) stated: "Clearly, the future of work will not be about college degrees; it will be about job skills."

In Australia, in February 2024, a major report into the future of Australia's higher education system proposed numerous systemic changes that align with the immersive STEM education and the training of career-ready graduates. These include initiatives such as establishing a tertiary education commission, which would have powers to manage university efforts to ensure skills of the right type and quantity can be provided to the nation's economy at any particular time. At a highly strategic level, this would help ensure that universities produce graduates that apply to the national economy.

However, in this document, concern was expressed about the manipulation of higher education fees to direct students towards more 'needed' graduate positions, and commentary in March 2024 (Matchett, 2024) suggests that universities have been too slow to implement innovative curricula and that Vocational Education & Training (VET) providers will receive a more significant number of student enrolments because of their typically more industry-focused, hands-on learning approach. Thus, there remains to be debate on the purpose of education and the need (or not) for it to provide career-ready graduates.

Further, Betts (2024), in an online forum (LinkedIn, 6 March 2024), suggested that the proposed reforms be revised to acknowledge the transformation in education to skills acquisition. Betts (2024) elaborates on this further and, as we suggest herein, suggests that higher education needs to change to be more flexible and skills/career-based so that life-long skill-based learning (active, immersive, hands-on learning) is more easily accomplished, enabling workers to upgrade their skills and competitiveness more dynamically.

Contemporaneously, countries such as Singapore acknowledge the requirement to specifically address industry demands and provide subsidies to workers under 40 to support their undertaking full-time study, enabling them to reskill for in-demand jobs (Chern, 2024).

9 Conclusion

In this chapter, we confirmed the desired attributes of a career-ready graduate before briefly examining the concepts of immersive learning. We then explored how such learning might occur, citing examples from the workplace, the laboratory or workshop, and the field.

Finally, considering recent practice and doctrine, we postulated how STEM education might be reimagined to produce better career-ready graduates. This is a dynamic area of discussion, and significant changes in how we prepare graduates are likely desirable in the immediate future if we are to educate graduates who can meet the rapidly evolving demands of industry.

References

Akour, M., & Alenezi, M. (2022). Higher education is in the future era of digital transformation. *Education Sciences, 12*(11), 784.

Andresen, L., Boud, D., & Cohen, R. (2020). Experience-based learning: Contemporary issues. In G. Foley (Ed.), *Understanding adult education and training* (2nd ed., pp. 225–239). Allen & Unwin.

Atkinson, J. (2012). What should geotechnical professionals be able to do? Shaking the foundations of geo-engineering education. McCabe, Pantazidou & Phillips (Eds.).

Atkinson, J. (2016). Basic geotechnical engineering skills: What can graduates do? In *International Society for Soil Mechanics and Geotechnical Engineering (ISSMGE), Shaping the Future of Geotechnical Education Conference, Brazil.*

Australian Army. (2018). Land Warfare Doctrine 7–6, Adventurous Training, Commonwealth of Australia (Australian Army).

Bandura, A. (2004). Observational learning. Learning and memory (2nd ed.). Ed. J. H. Byrne. Macmillan.

Bansal, G., Rajgopal, K., Chamola, V., Xiong, Z., & Niyato, D. (2022). Healthcare in metaverse: A survey on current metaverse applications in healthcare. *IEEE Access, 10*, 119914–119946.

Barraclough, J. (2016). *The resilience adventure, grounded curiosity (online blog).* http://groundedcuriosity.com/the-resilience-adventure/#.X0cNyS17H5Y

Berg, M. V., Paige, R. M., & Lou, K. H. (Eds.). (2023). *Student learning abroad: What our students are learning, what they are not, and what we can do about it.* Taylor & Francis.

Bogosian, B., Bobadilla, L., Alonso, M., Elias, A., Perez, G., Alhaffar, H., & Vassigh, S. (2020, March). Work in progress: Towards an immersive robotics training for the future of architecture, engineering, and construction workforce. In *2020 IEEE World Conference on Engineering Education (EDUNINE)* (pp. 1–4). IEEE.

Boyle, A., Maguire, S., Martin, A., Milsom, C., Nash, R., Rawlinson, S., & Conchie, S. (2007). Fieldwork is good: The student perception and the affective domain. *Journal of Geography in Higher Education, 31*(2), 299–317. https://doi.org/10.1080/03098260601063628

Campbell, I., & Price, R. (2016). Precarious work and precarious workers: Towards an improved conceptualization. *The Economic and Labour Relations Review, 27*(3), 314–332. https://doi.org/10.1177/1035304616652074C

Caten, C. S., Silva, D. S., Aguiar, R. B., Filho, L. C. P. S., & Huerta, J. M. P. (2019). Reshaping engineering learning to promote innovative entrepreneurial behavior. *Brazilian Journal of Operations & Production Management, 16*, 141–148.

Chern, K. W. (2024). Budget 2024: Businesses welcome upskilling benefits seen as vital for workers to stay relevant. *The Straits Times.*

Colburn, M. (2018). An alternative to categorizing skills as soft or hard. *OD Practitioner, 50*(4), 65–66.

Coomans, M. K. D., & Timmermans, H. J. P. (1997). Towards a taxonomy of virtual reality user interfaces. In *Proceedings of the International Conference on Information Visualisation (IV97)* (pp. 17–29).

Crawley, E., Malmqvist, J., Ostlund, S., Brodeur, D., & Edstrom, K. (2007). *Rethinking engineering education: The CDIO approach.* Springer.

Cunningham, W. V., & Villaseñor, P. (2016). Employer voices, employer demands, and implications for public skills development policy connecting the labour and education sectors. *World Bank Research Observer, 31*, 102–134.

Davis, G. (2017). *The Australian Idea of a University.* Melbourne University Press. Betts, 2024. Online forum, LinkedIn.

De Freitas, S., & Neumann, T. (2009). The use of 'exploratory learning' for supporting immersive learning in virtual environments. *Computers & Education, 52*(2), 343–352.

Dewey, J. (1938). *Experience and education.* Macmillan.

Diaz, C., Villeneuve, M., Penman, M., Brentnall, J., & Barton, R. (2015). The role of simulated fieldwork to bridge theory and practice. In *Australian & New Zealand Association for Health Professional Educators (ANZAHPE)/Asian Medical Education Association (AMEA) 2015 Joint Conference.*

Dyulicheva, Y. Y., & Glazieva, A. O. (2022). Game-based learning with artificial intelligence and immersive technologies: an overview. In *CEUR Workshop Proceedings* (Vol. 3077, pp. 146–159).

Engineers Australia. (2017). *Five essential soft skills for engineering graduates* (web page). https://yea.engineersaustralia.org.au/five-essential-soft-skills-engineering-graduates

Evans, N., & Acton, R. (2022). Narratives of teaching in outdoor and environmental education: What can we learn from a case study of outdoor education pedagogy? *Journal of Adventure Education and Outdoor Learning, 22*(3), 214–227. https://doi.org/10.1080/14729679.2021.1902828

Feinstein, A. H., Mann, S., & Corsun, D. (2002). Charting the experiential territory: Clarifying definitions and uses of computer simulation, games, and role play. *Journal of Management Development, 21*, 732–744. https://doi.org/10.1108/02621710210448011

Haywood, N., & Cairns, P. (2005). Engagement with an interactive museum exhibit. In T. McEwan, J. Gulliksen, & D. Benyon (Eds.), *People and Computers XIX, the bigger picture, proceedings of HCI 2005*. Springer.

Hermsen, P., Rooij, R., Rijnbeek, G., & Adrichem, T. (2022). *Reflection in engineering education* (White paper '100 DAYS OF… REFLECTION'). Delft University of Technology.

Hill, J., & Woodland, W. (2002) An evaluation of foreign fieldwork in promoting deep learning: A preliminary investigation. *Assessment and Evaluation in Higher Education, 27*, 539–555.

Hunt, L., & Chalmers, D. (Eds.). (2012). *University Teaching in Focus: A learning-centred approach* (1st ed.). Routledge. https://doi.org/10.4324/9780203079690

Jackson, D. (2018). Developing graduate career readiness in Australia: Shifting from extra-curricular internships to work-integrated learning. *International Journal of Work-Integrated Learning, 19*(1), 23–35.

Jaksa, M. B., Airey, D. W., Kodikara, J. K., Shahin, M. A., & Yuen, S. T. S. (2012). *Reinventing geotechnical engineering laboratory classes*. Shacking the Foundations of Geo-engineering Education. McCabe, Pantazidou, Phillips (Eds.). CRC Press.

Jennett, C., Cox, A. L., Cairns, P., Dhoparee, S., Epps, A., Tijs, T., & Walton, A. (2008). Measuring and defining the experience of immersion in games. *International Journal of Human-Computer Studies, 66*(9), 641–661. ISSN 1071–5819. https://doi.org/10.1016/j.ijhcs.2008.04.004

Jones-Vlasceanu, J. (2023). *AI's transformation of career preparation*. AACSB Insights. https://www.aacsb.edu/insights/articles/2023/07/ais-transformation-of-career-preparation

Karimi, H., & Pina, A. (2021). Strategically addressing the soft skills gap among STEM undergraduates. *Journal of Research in STEM Education, 7*(1), 21–46.

Kirk, B. (2021). *Adventure revolution: The life-changing power of choosing challenge*, Piatkus. McCallum, 2017. Personal Communication.

Kolb, A. Y., & Kolb, D. A. (2005). Learning styles and learning spaces: Enhancing experiential learning in higher education. *Academy of Management Learning and Education, 4*(2), 193–212.

Kolb, D. A. (2015). *Experiential learning: Experience as the source of learning and development*. Pearson.

Lauria, R. M., & Morie, J. F. (2015). VR as meta media and herald of our future realities, Chapter in Communication and Technology, L. Cantoni and J. A. Danowski (Eds.). Walter de Gruyter GmbH & Co KG.

Mandal & Edwards. (2022). *Student work readiness in Australian engineering workplaces through work integrated learning*. https://eric.ed.gov/?id=EJ1324787

Markowitz, D. M., Laha, R., Perone, B. P., Pea, R. D., & Bailenson, J. N. (2018, November 30). Immersive virtual reality field trips facilitate learning about climate change [Original Research]. *Frontiers in Psychology, 9*. https://doi.org/10.3389/fpsyg.2018.02364

Matchett, S. (2024). *How VET could eat universities' lunch*. Future Campus. https://futurecampus.com.au/2024/03/15/how-vet-could-eat-universities-lunch/

McCallum. (2017). *Personal Communication*, India.

McCallum. (2018). *Is the Future of Australian Higher Education Private?* Meanjin. https://meanjin.com.au/latest/isthe-future-of-australian-higher-education-private/

McCallum, A. B. (2020). *Postulations on empowering and enabling people to lead fulfilling lives...*, Linked-In. https://www.linkedin.com/pulse/postulations-empowering-andenabling-people-lead-lives-adrian-mccallum/

McCallum, A. B., & Fairweather, H. (2019). Empowering self-reflection to stimulate optimum outcomes in first year engineering. In *Australasian Association for Engineering Education Conference* (30th: 20189: Brisbane, Queensland).

McCallum, A. B., Porter, M., & Fairweather, H. (2016). Design, construction and investigation of an earth retaining structure—A foundation for active learning. In *AAEE 2016 Conference*, Coffs Harbour, Australia.

McKenzie, M. (2003). Beyond "The outward bound process": Rethinking student learning. *The Journal of Experiential Education, 26*(1), 8–23.

Mills, J. E., & Treagust, D. F. (2003). Engineering education—Is problem-based or project-based learning the answer? *Australasian Journal of Engineering Education.* The Australasian Association for Engineering Education Inc.

Mostafa, A., Elsayed, A., Ahmed, M., Mohamed, R., Adel, M., & Ashraf, Y. (2020). *Smart educational game based on augmented reality.* EasyChair Preprint no. 2731, EasyChair.

Munge, B., Thomas, G., & Heck, D. (2018). Outdoor fieldwork in higher education: Learning from multidisciplinary experience. *Journal of Experiential Education, 41*(1), 39–53. https://doi.org/10.1177/1053825917742

Munge, B., & Thomas, G. (2021). Managing outdoor fieldwork. In G. J. Thomas, J. Dyment, & H. Prince (Eds.), *Outdoor environmental education in higher education: International perspectives* (pp. 389–398). Springer.

Prikshat, V., Montague, A., Connell, J., & Burgess, J. (2020). Australian graduates' work readiness—Deficiencies, causes and potential solutions. *Higher Education, Skills and Work-Based Learning, 10*(2), 369–386. https://doi.org/10.1108/HESWBL-02-2019-0025

Ribeiro, R. (2012). Levels of immersion, tacit knowledge and expertise. *Phenomenology and the Cognitive Sciences. 12.* https://doi.org/10.1007/s11097-012-9257-z

Römgens, I., Scoupe, R., & Beausaert, S. (2020). Unravelling the concept of employability, bringing together research on employability in higher education and the workplace. *Studies in Higher Education, 45*(12), 2588–2603.

Ruhanen, L. M. (2006). The application of experiential learning techniques to event management education. *Global Events Congress and Event Educators' Forum, Customs House, Brisbane, Australia*, 27–29 September, 2006. The University of Queensland Press.

Schaffer, V. (2017). Enhancing learning to diverse cohorts via immersive visualization. *Journal of Hospitality, Leisure, Sport & Tourism Education, 21*(Part A), 46–54. ISSN 1473-8376. https://doi.org/10.1016/j.jhlste.2017.07.001

Sheedy, C. (2023). *Why candidates with soft skills are in such demand. Talent acquisition.* SEEK employer. https://www.seek.com.au/employer/hiring-advice/candidates-soft-skills-demand

Shulman, L. S. (2005). Signature pedagogies in the professions. *Daedalus, 134*(3), 52–59.

Testa, G. (2024). *What is career readiness in STEM? Defining the concept and its importance.* Labsta. https://www.labster.com/blog/career-readiness-stem-concept-important

Thomas, G. (2015). Signature pedagogies in outdoor education. *Asia-Pacific Journal of Health, Sport and Physical Education, 6*, 113–126.

Torisu, T. (2016). Sense of presence in social VR experience. *Interactive Architecture Laboratory.* UCL Bartlett School of Architecture.

Verner, I., Cuperman, D., Perez-Villalobos, H., Polishuk, A., & Gamer S. (2022). Augmented and virtual reality experiences for learning robotics and training integrative thinking skills. *Robotics, 11*(5), 90. https://doi.org/10.3390/robotics11050090

Vygotsky, L. S. (1978). *Mind in society: Development of higher psychological processes* (M. Cole, V. Jolm- Steiner, S. Scribner, & E. Souberman, Eds.). Harvard University Press. https://doi.org/10.2307/j.ctvjf9vz4

Westera, R., Prada, S., Mascarenhas, P. A., Santos, J., Dias, M., Guimarães, K., Georgiadis, E., Nyamsuren, K., Bahreini, Z., Yumak, C., Christyowidiasmoro, M., Dascalu, G., & Gutu-Robu, S. R. (2020). Artificial intelligence moving serious gaming: Presenting reusable game AI components. *Education and Information Technologies, 25*, 351–380. https://doi.org/10.1007/s10639-019-09968-2

World Economic Forum. (2020). In R. Doell (2020) Is a university degree necessary for future success? *Times Higher Education.* https://www.timeshighereducation.com/counsellor/counselling/university-degree-necessary-future-success

Zhao, X., Ren, Y., & Cheah, K. S. L. (2023). Leading Virtual Reality (VR) and Augmented Reality (AR) in education: Bibliometric and content analysis from the web of science (2018–2022). *Sage Open, 13*(3). https://doi.org/10.1177/21582440231190821

Chapter 6
Preparing Doctoral Program Graduates for Effective Academic Teaching and Disseminating Best Teaching Practices via the Scholarship of Teaching and Learning (SoTL)

Lisa Bosman and Rhea Dutta

Abstract The Scholarship of Teaching and Learning (SoTL) refers to a research and inquiry approach within higher education that focuses on studying teaching and learning processes to enhance educational practices and student outcomes. SoTL involves systematic investigation, reflection, and dissemination of findings related to teaching strategies, curriculum design, assessment methods, and other aspects of the teaching and learning environment. With SoTL, the goal is to improve teaching effectiveness and promote student learning through evidence-based practices. Unfortunately, many disciplinary engineering PhD programs, excluding Engineering Education, do not adequately prepare doctoral program graduates for effective teaching and dissemination of best teaching practices via SoTL. As we look to broaden participation in engineering, the limited scholarly training of disciplinary engineering educators causes a potential challenge for implementing effective pedagogical approaches critical for improving representation among underrepresented populations. The purpose of this book chapter is to provide readers with tools and strategies to implement innovative curriculum in the engineering classroom and share lessons learned via SoTL-focused journals and conference proceedings.

1 Introduction

The purpose of doctoral programs in multifaceted and can very depending on the field of study and specific goals of the program (Cardoso et al., 2022; Sarrico, 2022), including but not limited to preparation for academic positions, working in research and development in an industry position, advanced specialization, preparation for leadership roles, etc.... Yet, for most, doctoral programs are the gateway to academic

L. Bosman (✉) · R. Dutta
Purdue University, West Lafayette, IN, USA
e-mail: lbosman@purdue.edu

I. Gratchev et al. (eds.), *Transformative Approaches to Career-Ready Education*,
https://doi.org/10.1007/978-981-96-7018-5_6

positions, aiming to prepare candidates with the depth of knowledge and research skills necessary for a successful career in teaching and scholarship (Golde, 2004; Sherman et al., 2021). However, several issues with doctoral training have been identified, reflecting broader challenges within academia (Campbell et al., 2005). These issues can affect the readiness and ability of new Ph.D. graduates to effectively transition into academic teaching roles.

In general, doctoral programs tend to have an overemphasis on research (Ku et al., 2008; MacMillan et al., 2023). Doctoral candidates often receive minimal formal training in pedagogy, course design, assessment methods, and classroom management, which are crucial for a successful teaching career (Coso Strong & Sekayi, 2018). Although there is a recent trend towards teaching preparedness (Bonner et al., 2020), many doctoral programs prioritize research skills and outputs over teaching preparation. This can leave graduates well-prepared for research roles but underprepared for teaching responsibilities (Bok, 2013).

This overemphasis on research is often explained by academic job market saturation (Ball et al., 2015; Passaretta et al., 2019). There is a significant imbalance between the number of doctoral graduates and the availability of tenure-track positions, leading to high competition and precarious short-term or adjunct positions that offer little security or opportunity for career development (Ganning, 2022). As such, the skills prioritized within doctoral programs may not align with those required for available teaching positions, especially in institutions that value teaching over research (Boden & Nedeva, 2010; Golde & Dore, 2001; Kyvik & Olsen, 2012).

The purpose of this book chapter is to provide readers with tools and strategies to implement innovative curriculum in the engineering classroom and share lessons learned via SoTL-focused journals and conference proceedings. The next section provides an overview of SoTL. The following sections go into details on tools and strategies to become engaged with SoTL.

- Tool #1: Teaching Blueprint
- Tool #2: Peer Feedback Tuning Protocol
- Tool #3: Assessment of Student Learning
- Tool #4: Curriculum Intervention Summary
- Tool #5: Virtual Writing Groups
- Tool #6: SoTL Template

The book chapter concludes with supplemental resources for getting involved with SoTL.

2 Scholarship of Teaching and Learning

2.1 Definitions of SoTL

SoTL stands for the "Scholarship of Teaching and Learning." It's a movement in higher education that focuses on research and inquiry into the effectiveness of teaching practices and student learning outcomes (Booth & Woollacott, 2018; Gilpin & Liston, 2009). The aim of SoTL is to enhance teaching and learning through evidence-based methods, by conducting rigorous research on teaching methodologies, curriculum design, assessment techniques, and other aspects of the teaching–learning process. SoTL encourages faculty members to engage in scholarly inquiry into their own teaching practices and to share their findings with the broader academic community to improve teaching effectiveness and student learning experiences.

2.2 Benefits of SoTL

SoTL has many benefits (Haigh, 2012; Poole & Chick, 2022; Wahman et al., 2020).

First, employing SoTL encourages educators to reflect on their teaching methods and experiment with new strategies based on evidence. This can contribute to faculty members' professional development by fostering a deeper understanding of teaching principles, educational theories, and research methodologies. Moreover, this can lead to more effective teaching practices that better engage students and promote learning, and lead to greater job satisfaction and fulfillment among educators.

Second, engaging in SoTL enhances student learning. By systematically studying teaching and learning processes, educators can identify approaches that are most effective in facilitating student understanding and retention of material. This can result in improved learning outcomes for students.

Third, practicing SoTL provides promotion of evidence-based practices. SoTL encourages the use of evidence-based practices in teaching and learning. By conducting research on teaching methods and sharing findings with colleagues, educators contribute to a culture of evidence-based decision-making in academia. In addition, SoTL allows for contributions to the scholarship within the discipline. SoTL research adds to the body of knowledge within a discipline by exploring how students learn specific subject matter and how teaching methods impact learning outcomes. This contributes to the ongoing development of pedagogical theory and practice within the discipline.

Overall, SoTL offers a systematic approach to enhancing teaching effectiveness, improving student learning outcomes, and advancing the scholarship of teaching within the academic community.

2.3 Challenges of SoTL

Engaging in SoTL can also come with challenges (Fukuzawa et al., 2020; Kim et al., 2021; Vorster, 2020).

First, there are time constraints. Engaging in SoTL requires significant time and effort for designing research studies, collecting and analyzing data, and disseminating findings. Many faculty members already face heavy workloads due to teaching, research, and service responsibilities, making it difficult to dedicate time to SoTL activities.

Second, training and expertise is limited, as well as often a lack of institutional support. Conducting high-quality SoTL research requires knowledge of research methodologies, data analysis techniques, and ethical considerations specific to the field of education. Many faculty members may lack formal training in educational research, making it challenging to design and execute rigorous SoTL studies. Some institutions may not provide adequate support or recognition for SoTL efforts, which can discourage faculty members from engaging in this type of scholarship. Without institutional support in the form of funding, resources, and incentives, SoTL initiatives may struggle to thrive.

Third, for those new to SoTL, considering assessment and ethics may seem daunting. Assessing teaching effectiveness and student learning outcomes in a rigorous and meaningful way can be complex. Identifying appropriate assessment measures, collecting relevant data, and interpreting results accurately require careful consideration and expertise in assessment principles. In additional, SoTL research involving human participants raises ethical considerations related to informed consent, confidentiality, and potential risks to participants. Faculty members must navigate these ethical issues responsibly and adhere to institutional research ethics guidelines.

In summary, addressing these challenges requires a concerted effort from educators, institutions, and the broader academic community to provide support, training, resources, and recognition for SoTL efforts, ultimately promoting excellence in teaching and learning.

2.4 Current Approaches (and Gaps) to SoTL Professional Development

Current approaches to SoTL professional development vary depending on institutional priorities, resources, and the needs of faculty members. Some common approaches and strategies include (1) workshops and seminars, (2) faculty learning communities, (3) mentorship programs, (4) grants and funding opportunities, and (5) online resources and communities. Yet, gaps exist, primarily due to limited institutional support, lack of training and expertise, dissemination challenges, assessment complexity, and uncertainty related to institutional review boards and ethics considerations. Addressing these gaps and challenges requires collaborative efforts from institutions, faculty development professionals, and the broader academic community to provide support, resources, and recognition for SoTL professional development initiatives. By investing in SoTL professional development, institutions can promote excellence in teaching and learning and enhance the scholarly contributions of faculty members to the field of education.

3 SoTL Tools

3.1 Teaching Blueprint

The Teaching Blueprint (Bosman & Fernhaber, 2018) offers a mad-libs like canvas and checklist which leverages backward course design. Thus, it offers two main benefits. First, it provides a space for educators to jot down ideas. Second, it uses backward course design to promote alignment between the learning goal, learning objective, learning activity, and learning assessment. Backward course design, also known as backward design or backward planning, is an instructional design approach that starts with the end goal of learning in mind before planning the instructional activities.

Applying backward course design has many benefits (Davidovitch, 2013; Reynolds & Kearns, 2017; Wiggins & McTighe, 2005; Wilson, 2023).

1. Encourages Alignment with Learning Outcomes: Backward design ensures that course content, activities, and assessments are aligned with the intended learning outcomes or objectives. By starting with the end goals, instructors can design instruction that directly supports the desired student learning outcomes.
2. Reinforces Learning: Backward course design keeps the focus on what students will be able to do or understand at the end of the course rather than simply covering content. This approach emphasizes meaningful learning experiences and encourages instructors to prioritize essential knowledge and skills.
3. Endorses Clarity and Transparency: Backward course design requires clearly articulating learning outcomes upfront, which allows instructors to be more effective at communicating expectations to students, helping them understand what they are supposed to achieve by the end of the module or course.
4. Promotes Assessment Integration: Assessments are integral components of backward design. By identifying desired learning outcomes first, instructors can develop assessments that accurately measure students' achievement of those outcomes. This ensures that assessments are meaningful and relevant to the course objectives.
5. Offers Flexibility and Adaptability in Instructional Planning: Since instructors have a clear understanding of the desired outcomes, they can adjust instructional

strategies and activities as needed to better support student learning throughout the course.

6. Advocates for Higher Order Thinking: By designing instruction to support the achievement of specific learning outcomes, instructors can create opportunities for students to engage in meaningful and differentiated cognitive tasks, such as critical thinking, problem-solving, and application of knowledge.

In summary, backward course design provides a structured and systematic approach to course planning that prioritizes student learning outcomes, promotes alignment between course components, and enhances the overall effectiveness of teaching and learning experiences.

The Teaching Blueprint (Fig. 1) is a visual canvas that promotes backwards course design through alignment of the learning goal, learning objective, learning activity, and learning assessment. It highlights requirements for curriculum design essentials: (1) Key Teaching Intervention (e.g., combining the entrepreneurial mindset, STEAM, and bio-inspired design), (2) Professional Skill Development (with particular focus on communication and collaboration), (3) Mindset Cultivation (using multiple touchpoints and feedback to develop a habit, and (4) Integration of Best Teaching Practices.

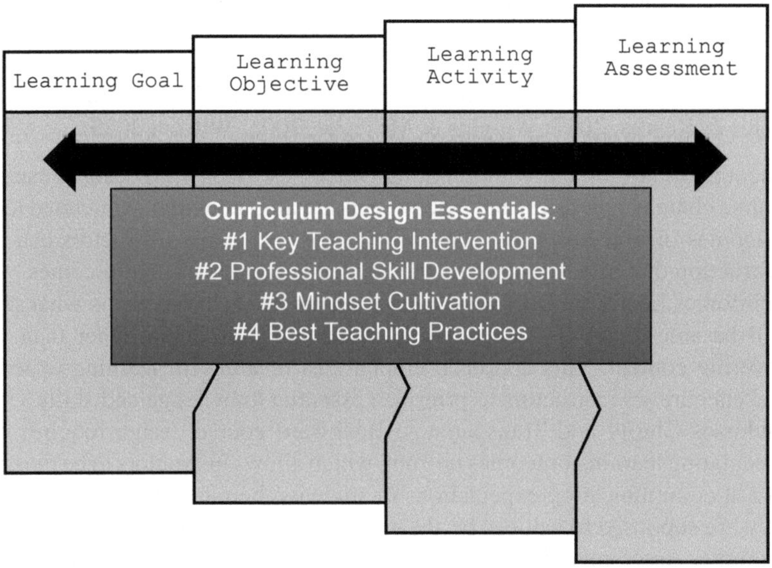

Fig. 1 Teaching blueprint

3.2 Peer Feedback Tuning Protocol

The Peer Feedback Tuning Protocol is a structured process for receiving peer feedback. It has two key benefits. First, it allows for two-way feedback and reflection. Second, it is time-controlled to ensure efficiency and equity in giving and receiving feedback. Peer feedback plays a crucial role in teaching new curriculum. This approach, where teachers give and receive feedback from their colleagues, enhances the teaching and learning process in various ways (Allen, 1995; Cushman, 1996; Gormally et al., 2014; Keiler et al., 2020; Paulsen et al., 2016; Zeng, 2020).

1. Promotes Diverse Perspectives: Instructors come with varied experiences and insights. When teaching new curriculum, these diverse perspectives can help identify potential challenges, strengths, and opportunities that might not be evident to a single educator. This can lead to a more robust and well-rounded approach to implementing the new curriculum.
2. Allows for Shared Understanding: Through peer feedback, professors can develop a shared understanding of the new curriculum's goals, content, and pedagogical strategies. This ensures a more consistent approach across different classes and grade levels, providing a coherent learning experience for students.
3. Encourages Personal Growth: Giving and receiving feedback is a powerful professional development tool. It encourages educators to reflect on their practice, consider new strategies, and continuously improve their teaching methods. This reflective practice is particularly important when adapting to a new curriculum, as it may require different approaches or methodologies.
4. Offers a Supportive Environment: Implementing a new curriculum can be daunting. Peer feedback creates a supportive network where educators can share challenges, solutions, and resources. This collegial support can reduce the sense of isolation and increase teachers' confidence in their ability to deliver the new content effectively.
5. Increased Engagement and Motivation: When educators are involved in the process of giving and receiving feedback, they feel more engaged and invested in the success of the new curriculum. This sense of ownership and involvement can enhance their motivation to implement the curriculum effectively and to continuously improve their teaching strategies.
6. Endorses Quality Assurance and Continuous Improvement: Peer feedback acts as a form of quality assurance, ensuring that the new curriculum is being implemented effectively and meets its intended goals. Continuous feedback loops allow for ongoing adjustments and refinements, leading to constantly improving educational offerings.
7. Opportunity to Model Reflective Practices: Peer feedback is a way for educators to model reflective practices for their students. When teachers engage in peer feedback, they model important skills such as critical thinking, constructive criticism, and reflective practice for their students. This not only benefits the teachers but also enriches the learning culture within the school, promoting a growth mindset among students.

In summary, peer feedback is a vital component of introducing a new curriculum, as it leverages the collective expertise, experiences, and support of the teaching staff to ensure a successful and dynamic educational experience for students. The Peer Feedback Tuning Protocol is a structured process for receiving peer feedback including timed steps. Many examples are available within the literature (Paulsen et al., 2016; Walsh, 2007).

3.3 Assessment of Student Learning

Assessment of student learning is a fundamental component of the educational process, serving multiple crucial functions that benefit both educators and students. Its importance can be outlined across several key areas (Dochy et al., 2007; El-Maaddawy, 2017; Finelli & Froyd, 2019; Hargreaves, 1997; Rompelman, 2000).

1. Important for Measuring Learning Outcomes: Assessment provides a tangible measure of what students have learned in relation to the curriculum's objectives. It helps in determining whether students have achieved the required understanding and skills in a subject area, facilitating the identification of both strengths and areas for improvement.
2. Critical for Guiding Instruction Decisions: The feedback from assessments allows educators to adjust their teaching strategies and content delivery. It helps in identifying topics that require further clarification or additional instructional time, ensuring that teaching methods are effectively meeting students' learning needs.
3. Crucial for Identifying Student Needs: Assessment helps in recognizing the diverse learning needs of students. By understanding individual performance, teachers can tailor their instruction to meet these needs, potentially through differentiation or personalized learning plans. This is crucial for supporting all students, including those who may require additional help and those who are ready for more challenging material.
4. Drives Improvements to Student Learning: Regular assessment encourages students to reflect on their own understanding and to identify areas where they need to improve. It promotes a proactive approach to learning, where students are not just passive recipients of information but are actively engaged in their own educational journey.
5. Motivates Students: Well-designed assessments can motivate students by setting clear goals and expectations. When students understand what is expected of them and receive feedback on their progress, they are more likely to be motivated to achieve those standards.
6. Promotes Institutional Accountability. Assessments provide evidence of student learning and teacher effectiveness, which is essential for accountability purposes. Schools and educators can use assessment data to demonstrate their success in

Photovoice Directions: *Please respond to the photovoice reflection prompts using three pictures (e.g., photo) and writing a narrative (e.g., voice). The narrative (minimum of 200 words for entire response) should reference the pictures and their relationship with respect to your response. Be sure to check assignment for spelling and grammar prior to submission.*

Example Photovoice Prompt (Technical Learning): Explain what technical content was learned by participating in the new curriculum.

Reflection Directions: *Please respond to the open-ended reflection questions with a minimum of 200 words per questions. Be sure to check assignment for spelling and grammar prior to submission.*

Example Reflection Prompt (Metacognitive Debrief): What went well? What didn't go so well? What will you do differently next time?

Fig. 2 Example assessment of student learning

meeting educational standards and to justify the need for additional resources or changes in instruction.

7. Facilitates Educational Research and Policy Making: Data from assessments can inform educational research and policy-making, offering insights into how different teaching methods, curricula, and policies affect student learning. This can lead to evidence-based decisions that enhance educational quality and outcomes.

In conclusion, assessment is not just about grading; it's a multifaceted tool that supports and enhances the learning process. It provides critical feedback to both students and teachers, guiding educational strategies, improving learning outcomes, and preparing students with the skills they need for future success.

One approach to the Assessment of Student Learning employs open-ended responses of photovoice and reflection. With the goal of exploratory analysis, open-ended questions can relate to (1) lessons learned, (2) likes and dis-likes about the new curriculum, (3) feedback on how the new curriculum connects theory to practice, and (4) a debrief about how they will leverage the experience and knowledge gained moving forward. Example prompts are provided in Fig. 2:

3.4 Curriculum Intervention Summary

SoTL involves systematic inquiry into teaching practices and student learning within a disciplinary context, with the aim of enhancing or improving the educational process. When disseminating findings, the methods section is crucial as it outlines how the study was conducted, ensuring that the research is replicable, verifiable, and understandable. The curriculum intervention summary is used to summarize information which belongs in the methods section.

1. Research Design: This could include whether the study is qualitative, quantitative, or mixed-methods. Here, it is important to provide justification for choice of research design, explaining how it is suitable for answering the research questions or achieving the study's objectives.
2. Study Context and Setting: This could include the type of institution (e.g., community college, university), the discipline or subject area, and any specific course or module details if relevant. The context provides background information that helps readers understand the environment in which the study took place.
3. Unidentifiable Demographics of Participants: This could include how the participants were selected and any relevant demographic information (e.g., year of study, major, gender). Also, it is important to mention the number of participants and any inclusion or exclusion criteria used. If the study involves specific groups (e.g., instructors, undergraduate students), details should be provided for each group.
4. Data Collection Methods: This could include specific instruments or tools used (e.g., surveys, interviews, observation checklists). If existing instruments were used, references should be provided. If new instruments were developed for the study, additional information should be provided with respect to validity and reliability. Also, a summary of how ethical issues were addressed should be included. This could include the process for obtaining approval from relevant ethics committees or institutional review boards; and state how participants were informed about the study, the consent process, and any measures taken to protect participant confidentiality and data security.
5. Data Analysis Protocols: This could include the specific statistical tests, coding schemes, or analytical frameworks applied to the data. For qualitative studies, a procedure should be used to explain how themes or patterns were identified. For quantitative studies, a reference to statistical software used and the rationale for choosing specific statistical methods should be included.

In conclusion, the curriculum intervention summary (e.g., info for the methods section) should be written with sufficient detail to allow another educator or researcher to replicate the study if desired. This not only contributes to the transparency and credibility of the study but also allows for the accumulation of knowledge through subsequent studies building upon your research.

The Curriculum Intervention Summary is used to summarize content which will be used for the Methods section of a SoTL-focused article. The following information is imperative for this section:

(1) Author(s) and author order,
(2) Tentative title
(3) Tentative dissemination outlet
(4) High level summary of new curriculum intervention,
(5) Description of technical context supplemented by personal insights (e.g., lessons learned) to provide a complete picture to others,
(6) Learning objectives,

(7) Supplemental documents including (but not limited to) assignment directions, rubrics, and learning activities.

3.5 Virtual Writing Groups

Virtual writing groups have gained prominence, especially with the increasing digitalization of academic and professional environments. These groups, where participants come together online to discuss and share their writing projects, offer several significant benefits, making them an important tool for writers across various disciplines and career stages. Here are some key reasons why virtual writing groups are important (Bosman et al., 2020; Bourgault et al., 2022; Li, 2023).

1. Offers Accessibility and Flexibility: Virtual writing groups break down geographical barriers, allowing individuals from different locations to participate without the need for physical travel. This accessibility makes it easier for people to find groups that match their specific interests or needs. Additionally, these groups often offer more flexibility in scheduling, accommodating participants across different time zones and with varying personal and professional commitment.
2. Promotes Accountability and Motivation: Writing can be a solitary and, at times, daunting task. Being part of a virtual writing group creates a sense of accountability, as members often set goals or deadlines and share their progress. Knowing that others are expecting updates can significantly boost motivation and productivity, helping writers to stay on track with their projects.
3. Encourages Diverse Perspectives and Feedback: Virtual groups bring together individuals from varied backgrounds, disciplines, and levels of experience. This diversity enriches the feedback process, offering multiple perspectives on a piece of writing. Constructive criticism from fellow writers can highlight issues that the author might not have noticed and suggest improvements, enhancing the quality of the work.
4. Provides Community and Support: Writing, particularly academic or professional writing, can feel isolated. Virtual writing groups foster a sense of community, providing emotional support and encouragement. Sharing experiences, challenges, and successes with peers can alleviate feelings of isolation and build a supportive network that values each member's success. Virtual writing groups often serve as a hub for sharing resources, such as writing tools, research materials, publishing opportunities, and information about workshops or webinars. Access to such resources can significantly aid in the writing and publication process.
5. Instills Skill Development and Learning: Participation in a virtual writing group can be a valuable learning experience. Members not only receive feedback on their work but also learn by reviewing others' writing. This can help in honing critical analysis skills, understanding different writing styles and methodologies, and staying updated with best practices in writing and research.

Table 1 Virtual writing group—structured approach

Round	Task	Time recommended (min)	The ask
1	Clarification	1–3	Any yes/no or one answer clarifying questions to clear up confusion?
2	Positive feedback	1–3	What is good about the paper and should be kept?
3	Author's request	10–20	Author requests specific feedback in 3–5 focus areas
4	Remaining feedback	5–10	Any additional feedback you'd like to offer?

6. Endorses Good Writing Habits: Regular meetings and deadlines set by virtual writing groups encourage members to make writing a consistent part of their routine. This regular practice is crucial for improving writing skills, completing projects, and advancing personal or professional goals.
7. Enhances Networking Opportunities. These groups provide opportunities to connect with like-minded individuals, expanding professional and academic networks. Networking can lead to collaborations, partnerships, or other opportunities that might not have arisen otherwise.

In conclusion, virtual writing groups are important because they offer a structured, supportive, and collaborative environment that can significantly enhance the writing process. By facilitating access to diverse perspectives, accountability, motivation, and a wealth of resources, these groups play a crucial role in the development and success of writers worldwide.

Virtual Writing Groups provide a structured process for receiving peer writing feedback (e.g., rounds that included clarification, positive feedback, responses to author's request, and any other feedback). Meetings can be held online. Writing is sent two days in advance for review. Feedback is shared in a structured manner during the 30-min online meeting (see Table 1).

3.6 SoTL Outline

A SoTL outline serves as a structured guide designed to assist educators in systematically investigating and reflecting on their teaching practices and the learning experiences of their students. The primary goal of SoTL is to enhance teaching effectiveness and student learning outcomes through scholarly inquiry. The use of a SoTL template can be particularly helpful for several reasons (Major & Braxton, 2020; Maurer, 2011; Mueller, 2018; O'Brien, 2008).

1. Provides Structure: A SoTL template offers a clear structure for conducting research, guiding educators through the necessary steps from identifying a

research question to collecting and analyzing data, and finally, reporting the findings. This structured approach ensures that the study is methodically planned and executed, enhancing its validity and reliability.

2. Facilitates Comprehensive Coverage and Clarity: Templates often include prompts or sections that cover all critical aspects of a research study, including literature review, methodology, data analysis, and ethical considerations. This comprehensive coverage helps ensure that researchers do not overlook essential components of their study, leading to more thorough and robust research. By following a template, educators can maintain a clear focus on their research objectives and questions. The template helps to delineate the study's scope, preventing scope creep and ensuring that the research remains aligned with its initial goals.
3. Promotes Reflective Practices: The process of following a SoTL template requires educators to reflect deeply on their teaching practices, the learning environment, and student outcomes. This reflective practice is a core component of SoTL, leading to insights and understandings that can inform future teaching and learning enhancements.
4. Offers a Documentation and Dissemination Strategy: A well-designed SoTL template not only aids in the research process but also in documenting and disseminating findings. The structured format can be easily adapted into articles for publication in educational journals or presentations at conferences, contributing to the broader educational community's knowledge and practices.
5. Guides Collaboration: When working in teams, a SoTL template provides a common framework that all members can follow. This facilitates effective collaboration, ensuring that all team members are aligned in their understanding of the research process and objectives.

In conclusion, a SoTL template is an invaluable tool for educators engaging in scholarly teaching and learning. It provides a systematic approach to investigating teaching practices and student learning, thereby supporting the continuous improvement of educational quality and outcomes.

The SoTL Template (Table 2) is an outline which included main headings (e.g., Intro, Literature Review, Methods, Results, Discussion, and Conclusion), sub-headings, and recommendations for citations.

4 Supplemental Resources

SoTL is an emerging field that emphasizes research into teaching practices and student learning outcomes in higher education. For educators interested in SoTL, there are numerous professional development resources available, designed to enhance their understanding and application of research-based teaching strategies. These resources range from journals and associations to workshops, webinars, and grants. Engagement with these resources can significantly contribute to an educator's professional development in the SoTL field, providing insights into effective teaching

Table 2 SoTL outline

Title
Abstract
1. Introduction [Entire Introduction should include at least 4–6 citations]
1.1 Problem Identification
1.2 Study Overview
2. Literature Review [Each topic section should include at least 4–6 citations]
2.1 Topic—Expand on Problem
2.2 Topic—Expand on Current Approaches and Gaps
2.3 Topic—Expand on Proposed Solution (Intervention and/or Assessment)
3. Methods
3.1 Study Design
3.2 Intervention
3.3 Participants
3.4 Data Collection Instrument(s)
3.5 Data Analysis Procedure(s)
4. Results
4.1 4.n [Summarize Findings in Logical Sequence]
5. Discussion
5.1 Respond to Research Question/Objective
5.2 Lessons Learned and Instructor Tips
6. Conclusion
6.1 Summarize Main Takeaway
6.2 Implications for Practitioners
6.3 Limitations and Future Research

practices and fostering a culture of continuous improvement in higher education. Here's a look at some of the key resources:

Journals

- Teaching and Learning Inquiry: https://journalhosting.ucalgary.ca/index.php/TLI/index
- Journal of Scholarship of Teaching and Learning (JoSoTL): https://scholarworks.iu.edu/journals/index.php/josotl/index
- College Teaching: https://www.tandfonline.com/journals/vcol20
- International Journal for the Scholarship of Teaching and Learning (IJ-SoTL): https://academics.georgiasouthern.edu/sotlgsu/journal/
- IEEE Transactions on Education: https://ieeexplore.ieee.org/xpl/RecentIssue.jsp?punumber=13

Associations and Organizations

- International Society for the Scholarship of Teaching and Learning (ISSOTL): https://issotl.com/
- The Society for Teaching and Learning in Higher Education (STLHE): https://www.stlhe.ca/
- Professional and Organizational Development (POD) Network in Higher Education: https://podnetwork.org/

Conferences and Communities

- ASEE: https://www.asee.org/
- WEEF: https://weef2024.org/
- AAEE: https://aaee.org/
- ISSOTL Annual Conference: https://issotl.com/issotl24/
- Lilly Conferences on College and University Teaching and Learning: https://www.lillyconferences.com/
- STLHE Annual Conference: https://www.stlhe.ca/conferences/2024-stlhe-conference/
- SoTL Commons Conference: https://academics.georgiasouthern.edu/sotlgsu/commons/

Online Resources

- SoTL Accelerator: www.SoTLAccelerator.com
- Scholarship of Teaching and Learning Annotated Literature Database: https://researchsotl.wordpress.com/
- Center for Engaged Learning: https://www.centerforengagedlearning.org/
- Engineering Unleashed: https://engineeringunleashed.com/
- Hopscotch 4-SoTL: https://hopscotchmodel.com/4-sotl/

5 Concluding Remarks

In summary, academic career preparation is crucial for doctoral students, especially in the area of scholarly teaching and learning. Many doctoral students pursue academic careers, which often involve teaching responsibilities. Preparation in pedagogy and instructional methods is critical to equip them with the necessary skills to effectively convey complex concepts to students and contribute to the literature with respect to teaching effectiveness. The purpose of this book chapter was to provide readers with tools and strategies to implement innovative curriculum in the engineering classroom and share lessons learned via SoTL-focused journals and conference proceedings.

References

Allen, D. (1995). The tuning protocol: A process for reflection. *Studies on exhibitions No, 15.*

Ball, D. M., Gleason, W., & Peterson, N. J. (2015). From all sides: Rethinking professionalization in a changing job market. *Pedagogy: Critical Approaches to Teaching Literature, Language, Composition, and Culture, 15*(1), 103–118.

Boden, R., & Nedeva, M. (2010). Employing discourse: Universities and graduate 'employability.' *Journal of Education Policy, 25*(1), 37–54.

Bok, D. (2013). We must prepare Ph. D. students for the complicated art of teaching. *The Chronicle of Higher Education* (11).

Bonner, R. L., Stone, C. B., Mittal, S., Phillips, W., & Utecht, R. L. (2020). Preparing academics to teach: Example of a structured method of preparing doctoral students in business programs to teach. *Journal of Management Education, 44*(4), 435–463.

Booth, S., & Woollacott, L. C. (2018). On the constitution of SoTL: Its domains and contexts. *Higher Education, 75*, 537–551.

Bosman, L., & Fernhaber, S. (2018). *Teaching the entrepreneurial mindset to engineers.* Springer International Publishing.

Bosman, L., McCave, E., Goldstein, M., & Chelberg, K. (2020). WIP: Virtual writing group participation: Surprises & unintended out-comes. In *American Society for Engineering Education Annual Conference and Exposition, Paper ID #29549.*

Bourgault, A. M., Galura, S. J., Kinchen, E. V., & Peach, B. C. (2022). Faculty writing accountability groups: A protocol for traditional and virtual settings. *Journal of Professional Nursing, 38*, 97–103.

Campbell, S. P., Fuller, A. K., & Patrick, D. A. (2005). Looking beyond research in doctoral education. *Frontiers in Ecology and the Environment, 3*(3), 153–160.

Cardoso, S., Santos, S., Diogo, S., Soares, D., & Carvalho, T. (2022). The transformation of doctoral education: A systematic literature review. *Higher Education, 84*(4), 885–908.

Coso Strong, A., & Sekayi, D. (2018). Exercising professional autonomy: Doctoral students' preparation for academic careers. *Studies in Graduate and Postdoctoral Education, 9*(2), 243–258. https://doi.org/10.1108/SGPE-D-18-00005

Cushman, K. (1996). Looking collaboratively at student work: An essential toolkit. *Horace, 13*(2), 1–12.

Davidovitch, N. (2013). Learning-centered teaching and backward course design from transferring knowledge to teaching skills. *Journal of International Education Research (JIER), 9*(4), 329–338.

Dochy, F., Segers, M., & Gijbels, D. (2007). Assessment engineering: Breaking down barriers between teaching and learning, and assessment. In *Rethinking assessment in higher education* (pp. 97–110). Routledge.

El-Maaddawy, T. (2017). Innovative assessment paradigm to enhance student learning in engineering education. *European Journal of Engineering Education, 42*(6), 1439–1454.

Finelli, C. J., & Froyd, J. E. (2019). Improving student learning in undergraduate engineering education by improving teaching and assessment. *Advances in Engineering Education.*

Fukuzawa, S., Ashbourne, D., & Rawle, F. (2020). Overcoming challenges to impactful SoTL. In *Evidence-based faculty development through the scholarship of teaching and learning (SoTL)* (pp. 366–387). IGI Global.

Ganning, J. (2022). Doctoral education and the academic job market in planning. *Journal of Planning Education and Research*, 0739456X211067271.

Gilpin, L. S., & Liston, D. (2009). Transformative education in the scholarship of teaching and learning: An analysis of SoTL literature. *International Journal for the Scholarship of Teaching and Learning, 3*(2), 11.

Golde, C. M. (2004). The responsibility of doctoral programs for the career preparation of future faculty. *Peer Review, 6*(3), 26–30.

Golde, C. M., & Dore, T. M. (2001). At cross purposes: What the experiences of today's doctoral students reveal about doctoral education.

Gormally, C., Evans, M., & Brickman, P. (2014). Feedback about teaching in higher ed: Neglected opportunities to promote change. *CBE—Life Sciences Education, 13*(2), 187–199.

Haigh, N. (2012). Sustaining and spreading the positive outcomes of SoTL projects: Issues, insights and strategies. *International Journal for Academic Development, 17*(1), 19–31.

Hargreaves, D. (1997). Student learning and assessment are inextricably linked. *European Journal of Engineering Education, 22*(4), 401–409.

Keiler, L. S., Diotti, R., Hudon, K., & Ransom, J. C. (2020). The role of feedback in teacher mentoring: How coaches, peers, and students affect teacher change. *Mentoring & Tutoring: Partnership in Learning, 28*(2), 126–155. https://doi.org/10.1080/13611267.2020.1749345

Kim, A. S., Popovic, C., Farrugia, L., Saleh, S. A., Maheux-Pelletier, G., & Frake-Mistak, M. (2021). On nurturing the emergent SoTL researcher: Responding to challenges and opportunities. *International Journal for Academic Development, 26*(2), 163–175.

Ku, H.-Y., Lahman, M. K., Yeh, H.-T., & Cheng, Y.-C. (2008). Into the academy: Preparing and mentoring international doctoral students. *Educational Technology Research and Development, 56*, 365–377.

Kyvik, S., & Olsen, T. B. (2012). The relevance of doctoral training in different labour markets. *Journal of Education and Work, 25*(2), 205–224.

Li, Y. (2023). The effect of online collaborative writing instruction on enhancing writing performance, writing motivation, and writing self-efficacy of Chinese EFL learners. *Frontiers in Psychology, 14*, 1165221. https://doi.org/10.3389/fpsyg.2023.1165221

MacMillan, T., Wang, D., Otuyelu, F., & Hill, A. (2023). Perceptions of doctoral programs' training and preparedness for academic careers. *Journal of Social Work Education and Practice, 8*(1), 19–29.

Major, C. H., & Braxton, J. M. (2020). SoTL in perspective: An inventory of the scholarship of teaching literature with recommendations for prospective authors. *Journal of the Professoriate, 11*(2).

Maurer, T. (2011). Reviewer essay: On publishing SoTL articles. *International Journal for the Scholarship of Teaching and Learning, 5*(1), 32.

Mueller, R. (2018). Ensuring design alignment in SoTL inquiry: Merging research purpose and methods. In *SoTL in Action* (pp. 53–61). Routledge.

O'Brien, M. (2008). Navigating the SoTL landscape: A compass, map and some tools for getting started. *International Journal for the Scholarship of Teaching and Learning, 2*(2), 15.

Passaretta, G., Trivellato, P., & Triventi, M. (2019). Between academia and labour market—The occupational outcomes of PhD graduates in a period of academic reforms and economic crisis. *Higher Education, 77*, 541–559.

Paulsen, T. H., Clark, T. K., & Anderson, R. G. (2016). Using the tuning protocol to generate peer feedback during student teaching lesson plan development. *Journal of Agricultural Education, 57*(3), 18–32.

Poole, G., & Chick, N. (2022). Great introspections: How and why SoTL looks inward. *Teaching & Learning Inquiry, 10*.

Reynolds, H. L., & Kearns, K. D. (2017). A planning tool for incorporating backward design, active learning, and authentic assessment in the college classroom. *College Teaching, 65*(1), 17–27. https://doi.org/10.1080/87567555.2016.1222575

Rompelman, O. (2000). Assessment of student learning: Evolution of objectives in engineering education and the consequences for assessment. *European Journal of Engineering Education, 25*(4), 339–350.

Sarrico, C. S. (2022). The expansion of doctoral education and the changing nature and purpose of the doctorate. *Higher Education, 84*(6), 1299–1315.

Sherman, D. K., Ortosky, L., Leong, S., Kello, C., & Hegarty, M. (2021). The changing landscape of doctoral education in science, technology, engineering, and mathematics: PhD students, faculty advisors, and preferences for varied career options. *Frontiers in Psychology, 12*, 711615.

Vorster, J.-A. (2020). SoTL: A mechanism for understanding and finding solutions to teaching and learning challenges. *Scholarship of Teaching and Learning in the South, 4*(2), 6–21.

Wahman, M. L., Peplow, A. B., Kumar, R., & Refaei, B. (2020). Benefits of using lesson study for SoTL, cross-disciplinary research, and assessment. *International Journal for the Scholarship of Teaching and Learning, 14*(1), 2.

Walsh, F. (2007). Using a tuning protocol: A journey of creating and sustaining professional conversations among teacher educators. *Northwest Journal of Teacher Education, 5*(1), 2.

Wiggins, G. P., & McTighe, J. (2005). *Understanding by design.* ASCD Publications.

Wilson, J. L. (2023). Using backward design to create a more diverse, equitable, and inclusive principles course. *The Journal of Economic Education, 54*(4), 440–452.

Zeng, L. M. (2020). Peer review of teaching in higher education: A systematic review of its impact on the professional development of university teachers from the teaching expertise perspective. *Educational Research Review, 31*(100333), 1–16.

Chapter 7
Mission Portfolio: Putting the Transfer into Transferable Skills

Gayle Brent, Kimberley Reis, Sarah-Jane Gregory, David Chen, and Simon Howell

Abstract The phrase *transferable skills* describes skills applicable across diverse contexts, especially within the workplace. However, the literal interpretation of skill transfer is often unclear to students. While educators assume skills acquired in one setting (university) can seamlessly transfer to others (the workplace), reports persist in a skills gap, suggesting graduates lack transferable skills and/or the confidence to apply them in novel situations. Addressing this requires an approach that empowers students to align the skills they have developed via life experiences with their professional identity. Traditional assessments are not best suited to capturing complex competencies, including employability, in a form that can be effectively evaluated. Students' ability to transfer their skills is more likely with learner-centred strategies that frame unfamiliar situations as learning opportunities. One solution is using portfolios centred on active reflection to help students evidence capabilities and develop a longitudinal view of their experiences aligned with their professional identity. For final-year students transitioning to the workplace via placements, portfolio assessments can build confidence to transfer academic skills to practical applications. This chapter showcases portfolio assessment embedded in three placement courses and a co-curricular program in the STEM disciplines, demonstrating how evidence-based articulation of skills through reflection enables valuable skill transfer to unfamiliar contexts.

G. Brent (✉) · K. Reis · S.-J. Gregory · D. Chen
Griffith University, Nathan, QLD, Australia
e-mail: gayle.brent@griffithuni.edu.au

S.-J. Gregory
CQUniversity, Bundaberg, QLD, Australia

S. Howell
Griffith University, Gold Coast, QLD, Australia

I. Gratchev et al. (eds.), *Transformative Approaches to Career-Ready Education*,
https://doi.org/10.1007/978-981-96-7018-5_7

1 Introduction

The ongoing debate over terminology for graduates' workplace skills has led to the widely used phrase *transferable skills* (sometimes called 'generic skills') to reference skills that are "applicable across several social, and especially employment contexts" (e.g. Clanchy & Ballard, 1995, p. 164). An emerging issue for consideration is whether the literal interpretation of this phrase (i.e. transferring skills) is clear to students. Educators often assume that the skills students acquire in one context—such as a university—can seamlessly transfer to other settings, including the workplace. However, reports of a skills gap persist (e.g. Prinsley & Baranyai, 2015; Sarkar et al., 2016), suggesting that graduates either lack the requisite transferable skills or the confidence to apply them in new situations. In either case, it is imperative to improve student outcomes in terms of developing their skills and/or confidence to transfer their skills and knowledge to unfamiliar contexts effectively. A multifaceted approach to employability and career readiness that empowers students to align their broad experiences and skills developed with their professional identity is required to achieve this.

Berger and Wild (2017) suggest that traditional paper-based assessment is not suitable for assessing ill-defined ('fuzzy') or 'complex' competencies (which includes a student's employability). de Blaquiére et al. (2019) emphasize the significance of coherent narratives that enable graduates to understand and articulate their employability. Multiple scholars espouse the use of portfolios (including reflection) to enable students to evidence capability (e.g. Baston, 2010) and to overtly engage with professional identity development (e.g. Carter, 2021). Baston (2010) quantifies that the development of a portfolio is a process that can take students from an *episodic* to a *longitudinal* interpretation of their capability—a concept that wholly aligns with and reaffirms the core construct of transference.

Bransford et al. (2000) assert that transference is more likely to occur in highly learner-centred contexts and where active learning strategies are employed, while Simatele (2015) espouses reflective practice for confident transference. Akkerman and Bakker (2011) explore *boundaries* as opportunities for learning, encompassing diverse (learning) sites that encourage participation, collaboration and transfer. They contend that the emphasis on bounded and singular domains of work has been challenged, and they introduce the term *boundary crossing* to describe situations where professionals enter unfamiliar situations they may feel unprepared to undertake. This concept has implications for final-year students on the cusp of transition from university to the workplace and, more specifically, for students undertaking work placement or practicum courses where they are required to adapt and adjust their skills to tackle novel situations in a workplace context. Further, it considers how learning and assessment design in final-year placement courses can help facilitate students' confidence to transfer skills to narrow the disparity between academic learning and practical application.

This chapter will showcase the utilization of portfolio assessment and reflective practice in three final-year work placement/practicum courses within the STEM disciplines and an associated co-curricular professional development program. Through case studies, we aim to demonstrate how assessment grounded in reflective practice and evidence-based articulation of skills enables the transfer of valuable skills to new contexts for the student.

1.1 Holistic Employability

Employability is a multifaceted concept that has long been a focal point in Higher Education. Variously referred to as 'employability', 'employability-based learning', 'career development learning', or 'career readiness', scholars have endeavoured to craft definitions that are simultaneously accessible to all stakeholders yet comprehensive enough to encompass the diverse range of experiences that might shape the development of a student's employability, including how they subsequently leverage these skills and attributes for transformative lifelong learning (e.g. Harvey, 2000). de Blaquiére and colleagues (2019) observe that in the quest to capture the complex meaning of 'employability', descriptions and definitions have nonetheless been heavily influenced by a discourse of skills (see also Reid et al., 2021). This is unsurprising given the ongoing need for students to evidence competency to align with employer recruitment processes. That is, to demonstrate what they can offer an employer by framing their experiences in terms of desirable skills (de Blaquiére et al., 2019) aligned to expected performance in the workplace. Despite this, de Blaquiére et al. (2019) and others (e.g. Harvey, 2000; Wenger, 1998) assert that to enable students to 'join the dots' of their career narrative, employability should be considered holistically as "ongoing, life-long, situated learning" (de Blaquiére et al., 2019, p. 16).

1.2 Situated Learning

Yorke and Knight's (2006, p. 5) USEM (Understanding, Skilful Practice, Efficacy Beliefs, Metacognition) model of employability encompasses the notion that skills are not developed in isolation. They highlight skilful practices in context, or *situated practice*, as a critical factor for learning throughout life, study and employment. Contexts for learning can (and should) include a broad range of activities that exist outside the formal curriculum. Cushing et al. (2019) observe that diverse life experiences are fertile ground for the development of 'intangible qualities and behavioural characteristics' that contribute to employability capability. However, they caution that these are difficult to assess. Jackson and Bridgstock (2021) emphasize that students expect meaningful opportunities to interact with industry and favour practical contexts where they can apply their theoretical knowledge and extend their learning beyond the classroom. However, they, too, comment on the inherent challenge of defining and gauging individual learning to determine the relative benefit of practical work-based experiences.

Reid et al. (2021) delve into the concept of 'meaning-making' derived from unstructured life experiences, emphasizing experiential learning theory to guide students to explore their employability. They argue that "making meaning from experiences leads to learning" (p. 100). Notably, Reid and colleagues distinguish between 'classroom' and 'work-integrated learning (WIL)' scenarios and the 'unstructured' learning encountered in informal and extracurricular contexts. Their observation implies that WIL is a fundamentally structured learning experience. This is at odds with Jackson and Bridgstock (2021), who emphasize the value of unstructured elements of WIL for developing students' self-awareness and confidence.

1.3 Exploring the Value of Unstructured Experiential Learning

Berger and Wild (2017) draw on Biggs and Tang's 2007 work to affirm that student learning takes place when they construct meaning from their activities, including authentic settings that are closely aligned with 'real work settings' (p. 433), which include Work Integrated Learning (WIL), placement and practicum courses. Carter (2021) consolidates the work of multiple scholars to define WIL as "an umbrella term describing the intentional practice of integrating traditional academic learning with applied learning for and in the workplace, with the intention of the student making connections between and deriving…meaning from both contexts" (p. 463). The structures of the three curricular case studies profiled in this chapter (Information, Communication and Technology (ICT) placement/project, Science WIL and Planning practicum) fit within this 'umbrella description', as does the co-curricular program offered in Griffith University's Sciences Group (Faculty) 'STEM Professionals'. Given the incompatibility of traditional methods of assessment to measure students' development in the crucial skill areas of creativity, flexibility, problem-solving and other 'wicked' competencies (Berger & Wild, 2017), authentic assessments delivered via university-developed co-curricular programs have strong potential to augment career development in STEM education.

Experiential learning is best described as learning through practical hands-on experiences (e.g. Spanjaard et al., 2018). Multiple scholars have noted that experiential learning encompasses 'life-wide' and 'life-long' learning concepts. That is, experiential learning approaches are more universal than traditional learning, and they allow for divergent experiences to contribute to learning through subjective interpretation that leads to "cognitive, perceptual and behavioural [skill] development" (Spanjaard et al., 2018, p. 165). In their works exploring the *Connectedness Learning Approach,* Bridgstock and Tippet (2019) posit that valuable learning is achieved through situated practice that is embedded in an informal framework and that these informal frameworks encompass behaviours and processes that are common in the workplace such as "expert modelling, mentoring, explicit instruction, advice and feedback" (p. 6). Workplaces, then, can be categorized as centres of unstructured learning. It logically follows that the challenges in establishing, gauging, and evaluating what students have learned in informal learning contexts also pertain to workplace experiences where the student will encounter unpredictable and unexpected situations to apply knowledge and skills beyond the scope of their formally linked university assessment (i.e. a placement or practicum project with a defined objective). As Reid and colleagues (2021) note, this captures the essence of experiential learning theory, which states that experience is both the basis and the stimulus for learning.

1.4 Is the Skills Gap a Lack of Transferability of Skills Gap?

There is a significant observation about students' and graduates' lack of preparedness for the workplace—often referred to as 'the skills gap'. Guárdia et al. (2021) frame this as employer dissatisfaction with the skills and competencies of graduates, and they underscore evidence in the literature of "crucial shortcomings in higher education" (p. 255) that contribute to this, namely, ongoing use of traditional teaching methodologies, and lack of emphasis on practical application of skills.

Noted above is the suggestion that students do not lack skills; they lack the ability to articulate and evidence their skills in line with employer expectations. If students cannot confidently identify their skills and knowledge, they may struggle to apply their skills, learned in the context of structured university experiences, to unstructured and unfamiliar workplace settings. Reid et al. (2021) observe that 'meaning-making' turns experience into learning. We posit that similar 'meaning-making', for example, through active and purposeful reflection throughout a student's degree, converts *learning* to *preparedness for the experience*. That is, if a student is guided—dare we say, expected?—to engage in reflective practice from their first year of study, they will have a greater capacity to identify the real, transferable 'value' of their structured, curricular learning experiences and future workplace experiences. 'Meaning-making' involves determining personal value and the significance of experiences, making emotionally laden experiences communicable. It is the process by which learners use language to articulate their experiences to themselves and others (Reid et al., 2021).

1.5 Pedagogies of Portfolio

Baston (2010) described portfolios as 'disruptive' technology that would anticipate the "learning and teaching models of the coming decades" (para. 2). He predicted that because portfolio development and management shifts learning into the sphere of the student, rather than the institution, all students would eventually have a portfolio of some type (although he likewise acknowledged that this 'profound change' would take time). He asserted the 'cohesion power of the portfolio' that would allow students to shift their learning from '*episodic*' to '*longitudinal*', and he observed that using portfolios to search, tag, aggregate, export, and display their learning from in-class, in curricular and through non-curricular experiences would empower students to manage their learning. These concepts are fundamental to what might be described as the *pedagogies of portfolio*—the foundation curricular aspects of embedded portfolio development combined with reflective practice that allow students to enact the "M" (metacognition) in the USEM model for employability (Yorke & Knight, 2006).

Guárdia et al. (2021) explore an *Active Blended Learning Approach* as a 'new paradigm' for higher education institutions that combines active learning methodologies with blended learning settings. They include portfolios in this category as portfolio learning encompasses a *sense-making activity* [emphasis added] in *appropriate learning environments* [emphasis added] that *engages students* in knowledge construction, reflection, and critique and enables the development of *learner autonomy* [emphasis added] (definition taken from the Institute of Learning & Teaching in Higher Education of the University of Northampton, n.d. cited in Guárdia et al., 2021).

Incorporating longitudinal portfolio assessments addresses the educational needs of student learning outlined by Akkerman and Van Eijck (2013), allowing students to engage in meaningful reflection on their abilities and knowledge developed and applied across diverse contexts. Akkerman and Van Eijck (2013) describe this approach as expanding traditional learning beyond a single domain to encompass the student as a 'whole person' with both 'inside' and 'outside' (school) experiences that are porous, complex and 'horizontal'. The expansion to the horizontal learning domain in the context of career development supports the practice of career storytelling and the crafting of personal narratives, which enable students to shape and communicate their professional identities in relation to their self-image and worldview, as elucidated by Polkinghorne (1988, cited in Guárdia et al., 2021).

1.6 Portfolio and Professional Identity

Building on the concept of portfolio learning, the discourse shifts to a broader contemplation of employability and identity formation. The essence of portfolio learning lies in the powerful potential for students to evolve their portfolios from mere collections of past work into dynamic self-portraits that capture their professional identity and foster future-oriented perspectives (Bennett et al., 2016). Reis et al. (2023) emphasize the significance of emotional intelligence in enhancing students' employability. They advocate for a holistic approach to employability development that encompasses cultivating a student's professional identity, suggesting that truly preparing students for the future transcends merely assembling a list of skills and is instead centred on nurturing their sense of self within the professional realm.

The dialogue about employability and identity is further enriched by Brown and Hesketh (2005), who highlight the importance of focusing on identity development as a cornerstone for employability, urging the creation of a 'narrative of employability' that should emerge from reflective practice. Jackson (2014) offers insights into what shapes a graduate's identity, including a sense of purpose, self-esteem, confidence, disciplinary knowledge, ongoing personal and professional development, and the agility to adapt skills across diverse situations. She equates graduate identity with professional identity, referencing Ewan's (1988) interpretation of professional identity as a self-conception that engenders feelings of adequacy and satisfaction in one's professional role. In this chapter, we embrace the nuanced perspective on professional identity espoused by Nadelson and colleagues (2017), who define it in terms of how graduate identity is situated within the professional sphere. This encompasses the competencies, understanding, convictions, methodologies, and ethical principles that define a profession.

1.7 Reflection for Transfer

John Dewey's (1998) widely cited insight, that learning emerges from reflecting on experience rather than from experience alone, underscores the crucial role of reflection in developing professional identity. Through intellectual and metacognitive practices, individuals learn to evaluate and adjust their actions based on past experiences, enabling the transfer of skills and knowledge from formal learning (academic) situations to unstructured learning environments that include the workplace and work placements. Cushing et al. (2019) and Reid et al. (2021) emphasize the role of informal learning and a deliberate, reflective process for deriving meaning from experiences that, in turn, support the emergence of professional identity and the articulation of career narratives. The necessity of reflective practice in skills transfer—emphasized by scholars like Bransford et al. (2000), Simatele (2015), and Akkerman and Bakker (2011)—is critical for applying learned skills in new contexts. The discussion in this chapter and the case studies anchored the conversation about 'employability' and 'skill transfer' to creating coherent, reflective narratives within portfolios, showcasing the evolution of professional identity and capabilities aligned with students' anticipated future careers.

1.8 Practical Implementation

It is widely recognized that initiating career development early in a student's academic journey is crucial, yet engaging them in this process presents a notable challenge within higher education (e.g., Perrone & Vickers, 2003, as mentioned in Choate et al., 2019). Faulkner et al. (2013) point to students' reluctance to leave their comfort zones as a critical reason for their minimal engagement in personal and professional development activities, with many preferring to preserve the *status quo* rather than seek growth opportunities. Bridgstock (2009) and Bridgstock et al. (2019) advocate for employability-based learning that is integrated, assessed, and spans the entire curriculum.

Since 2014, the Sciences Group of Griffith University, Queensland, Australia, has prioritized a comprehensive, assessed approach to career development. From 2017 onwards, the Group emphasized using portfolios within Work-Integrated Learning (WIL) projects, placements, and practicum courses. The following section introduces three case studies of embedded portfolio assessments and one integrated, co-curricular example linked to embedded employability-based learning tasks across various STEM disciplines within the Sciences Group.

2 Case Studies

2.1 Introduction to Case Studies

The following case studies are drawn from three final-year Work Integrated Learning (WIL), Placement and Practicum courses (subjects) and a co-curricular professional development program (STEM Professionals) within the Griffith Sciences Group (Faculty). The Group comprises the School of Information Technology (ICT), the School of Engineering and Built Environment (EBE) and the School of Environment and Science (ESC), with one case study provided from each of these schools in the disciplines of Information Technology, Science and Urban and Environmental Planning.

Each of the courses and the co-curricular program, STEM Professionals, uses the Personal Learning Platform *PebblePad* to deliver comprehensive learning resources to support and scaffold the students' portfolio development. The learning resources are visually represented in Fig. 1 and outlined in greater detail in Table 1. The resources include an introduction to reflective writing for employability, drawing from life experiences to evidence skills, creating artefacts, aligning experiences with competency frameworks, and style ideas for designing a portfolio.

The functionality of PebblePad has been extensively leveraged to provide opportunities for students to engage in interactive learning modules at their own pace before designing their portfolios. In the sample provided (Fig. 1), taken from the Planning Practicum course (4017ENV), a worksheet is associated with each required artefact in the final portfolio submission. In this course and the STEM Professionals program, PebblePad has also enabled a scalable process for providing individualized, meaningful feedback as students progress through the assessment/activities (for example, via: "Week 4, Week 6 and Week 12" submissions as in Fig. 1). In the Science WIL courses, students are also provided with feedback on their artefacts' early development, as outlined in the relevant case study. In the Planning and Science examples,

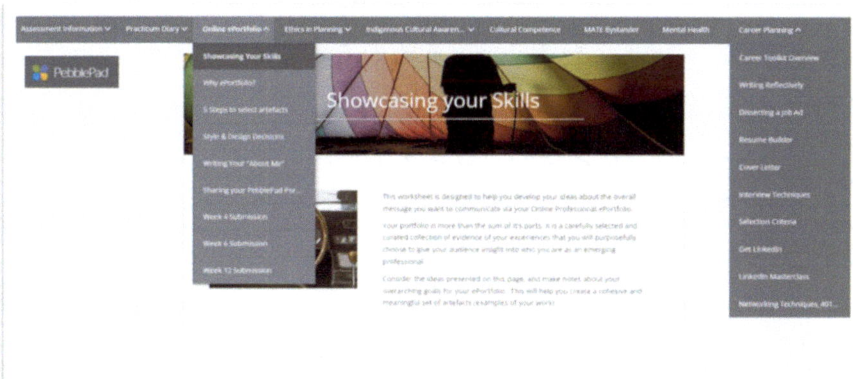

Fig. 1 An overview of learning resources provided via PebblePad to support portfolio development and reflective writing

the use of PebblePad has allowed for a cohesive approach to the portfolio development, which begins in the student's first year (in 1511ENV, the first year Studio, and 1015SCG Quantitative Reasoning (a core course), respectively) which is finalized in their final year.

3 The Planning Practicum: Case Study in the Fourth-Year Planning Degree

3.1 Overview

The Planning Practicum 4017ENV course is a significant milestone for fourth-year students in the Urban and Environmental Planning Program. It aims to introduce students to professional planning practice and equip them with the skills, knowledge, and tools required to enhance their employability after the degree program. Students must complete 217.5 hours of approved (paid or unpaid) work experience in a planning-related role and participate in a series of studio sessions exploring issues underpinning contemporary planning practice. Students in this course are considered to be 'early career professionals'; hence, the term 'Trainee Planners' refers to these students. The Learning Outcomes include:

1) Apply knowledge and skills acquired through the program to a professional planning work placement;
2) Evaluate and reflect on individual professional planning experience in light of academic and theoretical base, and vice-versa;
3) Expand and apply employability skills such as time management, interpersonal skills, professional development, communication and networking;

Table 1 Overview of portfolio modules in PebblePad workbook for courses/programs profiled in the case studies

Course/Program	How to write reflectively	What did you learn?	Competency Framework	Showcasing your skills	Creating Your Portfolio Why Portfolio I 5 steps to select artefacts I Style & Design Decisions I Writing your "About Me"	Professionalism / Ethics	Additional Modules
Science WIL Courses	✓	✓	✓	✓	✓		– Deliverables – Me-in-a-Minute
Planning Practicum Course	✓	✓	✓	✓	✓	✓	– Indigenous Competencies – Cultural Competence – MATE Bystander – Mental Health – Career Planning

(continued)

Table 1 (continued)

Course/Program	How to write reflectively	What did you learn?	Competency Framework	Showcasing your skills	Creating Your Portfolio Why Portfolio I 5 steps to select artefacts I Style & Design Decisions I Writing your “About Me”	Professionalism / Ethics	Additional Modules
ICT WIL Course	✓	✓				✓	– Track Your Learning (7822ICT) – Crafting Your Pitch
STEM Professionals	✓		✓	✓	✓	✓	– Career Toolkit Tips – Switch On: Opportunities – Teamwork – Career Conversations I Stories – Networking

4) Demonstrate professional awareness and contribute to the advancement of the planning profession;
5) Communicate knowledge of the discipline and professional practice to a professional audience - through written expression, oral presentations, multimedia, and
6) Effectively define and communicate appreciation for the planning profession and personal career aspirations to assist in the search for graduate employment.

3.2 Implementation of Portfolio Assessment

The overall 'Professional Portfolio' assessment includes two (2) components: an 'Online Professional Portfolio' and a reflective 'Practicum Diary', depicted in Fig. 2.

The 'Online Professional Portfolio' is an external-facing living showcase of individualized work examples with reflections on the development of relevant Planning Institute of Australia (PIA) graduate competencies, explicitly focusing on growing competencies in professional and ethical practice (PIA, 2020). These are written to a broader audience of prospective employers, which allows the Trainee Planner to tell their professional story. The assessment consists of the following elements:

a) An 'About Me' statement focuses on the Trainee Planner's motivations, why they are studying, and what they hope to achieve in the future. This statement includes a professional profile picture.
b) Three (3) compulsory examples of work and reflections to evidence the development of chosen PIA competencies that communicate a commitment to, skill in,

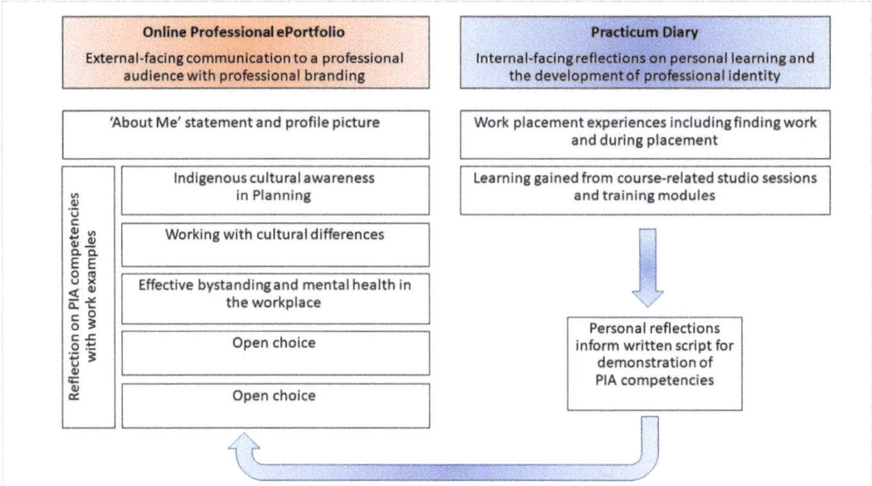

Fig. 2 Overview of professional portfolio task

or value for professional and ethical planning practice. These tasks help students to connect their learning with their emerging future as planning professionals.

c) Two (2) open-choice work examples (with reflection) to evidence capabilities and explain the link between the chosen experiences and the skills identified as important to a chosen PIA competency for professional and ethical planning practice.

The 'Practicum Diary' is an internal-facing reflective resource that is an important indicator of the Trainee Planner's 'inner commitment', that is, processes within oneself for developing reflective practice and how this informs professionalism and enhanced engagement. Baldwin and Rossier (2017) contend that the 'outer' life of the professional world is a meaningful expression, in the deepest sense, of the 'inner' values and insights of the progressively unfolding human experience. The Urban and Environmental Planning Program is professionally accredited by PIA every five years. This course complies with PIA's requirement that "reflective work-integrated learning that forms part of the educational curriculum to demonstrate how skills, knowledge and competencies obtained through work experience are acquired" (PIA, 2020, p. 33). The structure of the assessment consists of the following components:

a) Work placement reflections: Trainee Planners maintain records and reflections of work placement experiences, ranging over the pre-placement (work searching phase), during placement and post-placement experience. Trainee planners learn that record keeping is part of standard practice in the professional world, and these records additionally provide valuable information for consideration when preparing other assessment items for the course.
b) Studio Learning Reflections: Trainee Planners capture their learning from course-related studio sessions and training modules that support their interpretation of experiences within their placement. Their reflections are enhanced by scaffolded online modules delivered in PebblePad, focusing on a range of competencies and activities directly related to developing the compulsory artefacts for their Professional Portfolio.

The structure of the overall 'Professional Portfolio' assessment in the Planning Practicum 4017ENV builds upon the 'Professional Development Portfolio' that commences in the first-year Environmental Planning Studio 1511ENV. In the first year, the recording of student learning via the portfolio approach scaffolds the career development and industry engagement of students 'transitioning' to university with a focus on clarifying a sense of career aspiration (Reis et al., 2023). By the fourth year, the Planning Practicum portfolio scaffolds to the 'transitioning out' (Griffith University, 2024) phase, focusing on building a sense of mastery.

3.3 *Challenges, Outcomes and Evaluation*

Scaffolded learning for building a sense of mastery (Bell, 2024) through reflective practice, depicted in Fig. 3, is embedded within each year of the four-year planning

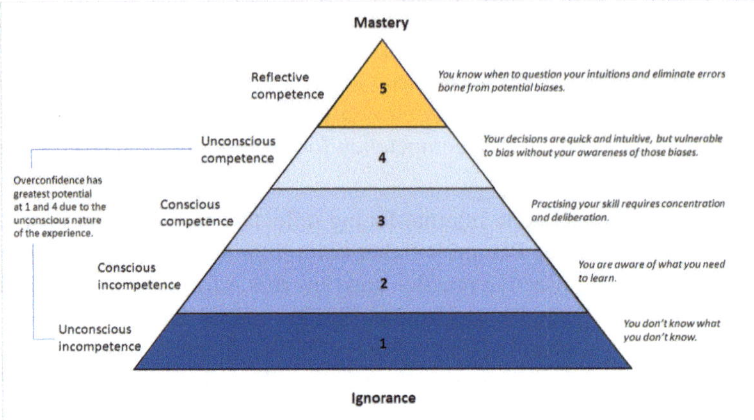

Fig. 3 Stages of Mastery (Adapted from Bell, 2024)

degree program. Over this time, students engage in various reflective activities that progress them from degrees of 'not-knowing' through repeated concentration and deliberation to generate more informed and, at times, intuitive insights.

The reflective components of the Practicum Diary and the Professional Portfolio assessments have had mixed reviews from Trainee Planners. A significant challenge for new graduates is the pervasive view in the planning profession that the time and reflection required to digest and interpret new insights is part of what makes academic publications seem inaccessible. Few planners with a practical dimension to their work make space for critical reflection of new insights (Taylor & Hurley, 2016), resulting in student perceptions that emerging planners are expected to focus exclusively on practical competence (Schon, 1983). As one student shared in the 2023 Student Experience of Teaching (SET) feedback process: "academia and real-world applications are two different things". Furthermore, sharing online portfolio reflections can be considered by Trainee Planners as superfluous to their career prospects. One student suggested via the 2023 Student Experience of Course (SEC) feedback "discontinuing the website task for PebblePad. [As they]...spoke to Directors of planning firms who said they never check [the] portfolios of their staff, so they did not understand the use of websites when LinkedIn is so available". As such, students do not see portfolio reflections on the questions of meaning and the value of loosening the tight boundaries around professional identity as a productive use of their time.

Other Trainee Planners found the assessment fulfilling. For example, one student shared that,

> Although I originally thought there were too many reflections, the content really helped me to better understand my placement experience, and I feel more confident in myself and my abilities. (SEC, 2022)

Another student reported that:

> Combining practicum with reflection and assessment really encourages an appreciation for the work we have completed during university and for the opportunity to take a step into professional practice. (SEC, 2023)

A third student highlighted the opportunity to question their intuitions and ongoing development, commenting that "this course has inspired me to be constantly thinking about how I can be a better planner" (NA-SEC-2023). A major driver of these insights is making space for shared discussions among Trainee Planners (within the Studio workshops) so they can debrief, learn from each other's experiences, and come to see their career narrative as a life-long and holistic journey (de Blaquiére et al., 2019).

4 Science Work Integrated Learning Undergraduate Courses

4.1 Course and Participant Overview

Within the School of Environment and Science, both undergraduate WIL courses have portfolio development as a major assessment component. WIL courses are pinnacle experiences that allow students to consolidate applied understanding of their discipline knowledge developed in university coursework and develop work-ready competencies. In the Science discipline, students generally take these courses in their final trimester and undertake either 12 or 24 full workdays across 12 weeks in an approved industry placement.

The portfolio is the key component of integrating the learning associated with the industry placement experience. Key learning outcomes include critically reflecting on experiences and curating evidence demonstrating work-ready capabilities and discipline competencies. Notably, the final portfolio is a crucial compilation of critical artefacts to showcase their career readiness to the industry.

4.2 Implementation of Portfolio Assessment

Rationale and competencies: The inclusion of portfolio assessment has been instrumental in ensuring the integration of learning and work experience; work experience and learning. It allows students to create and share a wide range of documentation and artefacts in varying forms for diverse audiences. Including early feedback provides an amplified opportunity for students to develop a strong awareness of the actual transfer of their skills to new contexts in the workplace. At the same time, the portfolio assessment is also intended to become a showcase portfolio that can be shared publicly as evidence of student excellence, experience, and capabilities. If students utilize the provided resources and feedback, it is anticipated that any student should be able to be highly succcssful in developing work of exceptional quality.

Connection to Industry Competency Framework: Many degree programs that include the WIL course do not have specific accreditation requirements from a governing professional body. Therefore, the portfolios require students to align their work with the Science Threshold Learning Outcomes (TLOs), which are a national reference point for science student competencies that align with a wide range of industry-desirable capabilities (https://www.acds.edu.au/teaching-learning/science-threshold-learning-outcomes-tlos/). The five critical areas of competency are broadly (i) Understanding Science, (ii) Scientific Knowledge, (iii) Inquiry and Problem Solving, (iv) Communication, and (v) Personal and Professional Responsibility.

Structure of the Portfolio Assessment: The portfolio assessment is divided into multiple sections - Part A, Part B and My-placement-in-a-minute (MPIAM) worth 10%, 50% and 15%, respectively (as shown in Fig. 4). The staged assessment approach allows students to receive feedback that supports excellence in their later submissions. Despite early introduction to portfolios in their first year, many of the tasks associated with portfolio development are new to the students as they are more specifically focused on immediate reflective practice within the work placement to identify the application and development of skills, coupled with retrospective reflective practice to evidence skills later in the trimester. Therefore, a range of support resources accompany each task to support students' learning and ability to demonstrate their capabilities within their portfolio effectively. Examples include a 'What is in it for me the student' statement (task rationale) about the task and how it will personally benefit the student, guidance statements on how each task can be approached, exemplars from previous students, and detailed marking rubrics. In addition, fortnightly lectures/workshops are held to help students further understand the assessment requirements and support their portfolio development. Figure 4 provides an overview of the staged submission structure of portfolio artefacts. At the same time, Table 2 gives extended detail on the rationale for each required element in the portfolio assessment, including how the artefact contributes to student learning.

4.3 Challenges

The implementation of portfolios has been met with some challenges as aspects of these assessment tasks are new to the students and are perceived to be less 'science' than other assignments and assignment types they have encountered and, therefore, less relevant. This is due to prior university assessment experiences, which have primarily focused on discipline knowledge and skills, with little, if any, embedded reflective practice. To address student concerns, the 'What is in it for me?' framework explaining the personal benefit to the students was developed and has helped change how students perceive the portfolio task. In addition, exemplars, detailed marking criteria, and individualized feedback on draft work have primarily overcome this challenge.

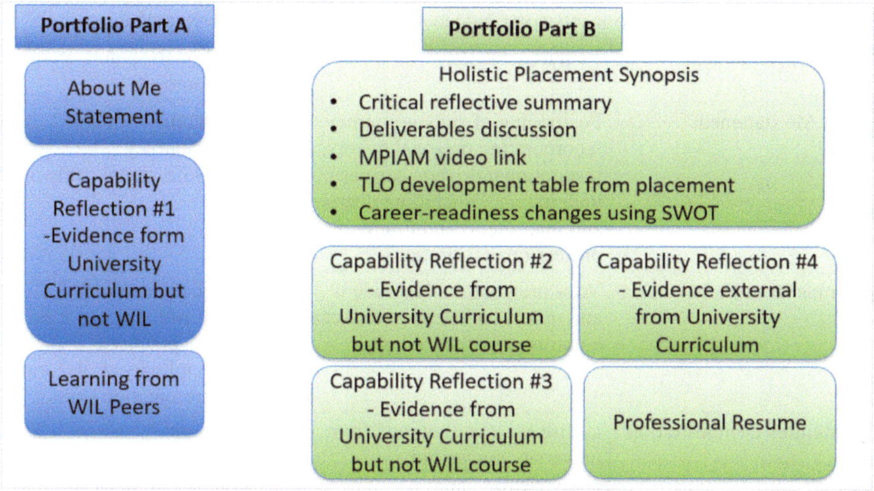

Fig. 4 Staged submission of portfolio artefacts in ESC WIL courses

4.4 Outcomes and Evaluation

Both students and industry partners have commented on the value of the portfolio task.

For students, a key outcome is the opportunity to be supported in developing a collection of artefacts that can be used in the search for a graduate role. Sample student comments include:

> The assessment items in this course all linked very well together and helped me create a portfolio I feel truly proud to show off. (SEC, 2023)
>
> I have been offered a graduate position already (not graduating until Nov!). They were really impressed with how I could apply my capabilities to their work context and the digital portfolio I developed in 3900ESC. (SEC, 2022)

Industry partners have commented similarly regarding their placement students' portfolios: "Holy smokes! That is incredible" (Industry partner, 2023). Additionally:

> This portfolio work is brilliant! There is so much good thinking, and the various bits are helpful in supporting students in putting their best foot forward professionally. It helps to make them very competitive candidates for future positions because they can clearly articulate what they can do. (Industry partner, 2024)

For universities, ensuring students can recognize and value what they have learnt during their degree enables them to apply their skills more confidently within an industry setting and contributes to their success during placement experiences. This confidence, in turn, increases students' opportunities to secure graduate roles and allows groups within the university to better support the transition from university to the workplace:

Table 2 How portfolio components contribute to student learning and development

Portfolio Component	Requirement	Contribution to student learning and development
About Me statement	Positioning of student journey incorporating personal history, inspirational drivers, current undertakings (including degree and placement) and future professional aspirations	Reflection on: • learning journey • future professional pathway
Learning from Peers	View previous student videos and reflect on learnings that can be utilised by the student	Communicate learnings associated with: • discipline practices, • transdisciplinary skills, oral/ visual presentation skills
Capabilities Demonstration	Utilise evidence sourced from university curriculum, co-curricular and extracurricular experiences to demonstrate Science TLOs,	Reflection on: • identification of sources that can be used to evidence capabilities, • effective storytelling Ability to: • apply a critical reflective writing framework, • effectively curate multiple sources of evidence to demonstrate capabilities
Holistic WIL Placement Experience Synopsis	This piece focuses specifically on their WIL Placement outcomes and includes: • Overview summary, • Science TLO skills development summary, • Discussion of Industry Deliverables, • Post-placement SWOT analysis and critical comparison to pre-placement SWOT for career-readiness	Critically reflection on their placement experience, how it has impacted both the student and their industry provider Recognition, articulation and evidencing of student produced industry deliverables and why these are valued by their industry partner Ability to implement feedback seeking strategies associated with discipline knowledges, professional capabilities, and quality of work
Resume	2-page resume	Effective resume curation Identification of skills and capabilities to value and highlight

(continued)

Table 2 (continued)

Portfolio Component	Requirement	Contribution to student learning and development
My-placement-in-a-minute	1 min video that summaries key aspects of their placement experience including: • Placement summary • Activities undertaken • Industry outcomes • Student outcomes • Placement impact on employability/ career-readiness	Video production skills Oral presentation skills Highlight curation of information
Overall portfolio	Requirement for an engaging digital showcase portfolio that utilises a variety of evidence and presentation strategies such as video, audio, critical reflection	• reflection and communication of skills, capabilities, and experiences for a variety of different future scenarios (e.g. job application, job interview, LinkedIn profile etc.) • write succinctly, • write an engaging and effective narrative, • curation of a professional identity and storytelling capacity, • utilisation of a wider range of evidence including laboratory reports, assignments, presentations, photographs, data, and importantly feedback on their capabilities from external stakeholders, • present themselves in multiple forms for different audiences, • digital portfolio curation skills

> Having students complete this kind of portfolio work makes it much easier for us to support students effectively. (Careers & Employment Services, 2022)

The implementation of portfolios for students undertaking WIL is highly beneficial. Portfolio best practice includes critical reflection on discipline skills, transferable skills, and knowledge aligned with placement learning and the development of suitable artefacts to ensure graduates can articulate their capabilities to different audiences. Students may not possess the skills to develop such portfolios, and thus, suitable support resources should be included in the curriculum design.

5 Bachelor of Information Technology Work Integrated Learning Placement Course

5.1 *Course and Participant Overview*

The School of ICT offers a 20CP WIL Placement capstone course in all its undergraduate and postgraduate programs. Students undertaking this course must be in the final year of their study and conduct a 12-week (two days per week) industry-based placement. The primary learning outcome is for students to apply a wide range of knowledge and skills they have developed in earlier courses within their degree program to a practical, industry-based body of work related to their major. Students must submit a series of project deliverables and a comprehensive reflective report on their experiences and learnings encapsulated in their portfolio.

5.2 *Implementation of Portfolio Assessment*

Historically, there have been significant challenges in effectively capturing student learning and assessing student achievement against the learning outcomes in iterations of the WIL placement course (3822ICT/7822ICT). The challenges arose from the very nature of the placement course in that:

1. Work undertaken during their placement is different for every student;
2. Students learn both ICT-specific skills and context-driven transferable skills in the course;
3. The course needs to assess both the depth and breadth of the student's learning;
4. Assessment needs to be at Bloom's Taxonomy Level 4 (or higher), centring on students' ability to analyze their experiences to establish a holistic view of their learning.

A further issue with the WIL course was that the students tended to focus on the practical work/project at the expense of reflecting on and documenting their learning. This translated to an overemphasis on placement deliverables, with students giving little attention to analyzing their decision-making, considering the impacts of their choices and evidencing their learning and skill development during their placement.

To address these challenges, a portfolio assessment was introduced in 2022, weighted at 40%, to ensure students have the opportunity and incentive to actively reflect on their experiences and compile evidence of their learning by writing a series of critical reflections on their placement experience. The portfolio task ensures students have sufficient guidance to meet all four learning objectives of the WIL course:

1. Successfully apply and further develop the knowledge and skills gained through your academic program professionally.

2. Work and learn independently as an effective IT professional.
3. Demonstrate a high level of ethical and professional behaviour.
4. Prepare comprehensive and appropriate documentation and presentations.

An overview of the portfolio assessment task and relevant learning outcomes is provided in Table 3.

Each reflection requires students to consider what they did well (developing confidence) and how they can improve and plan to do better (lifelong learning). Students are encouraged to keep a learning log throughout their placement experience to capture learning when it happens, although this is not a formal requirement of the portfolio assessment at this stage. The portfolio assessment is also designed to ensure students can articulate evidence of skills and knowledge they consider essential for their future careers. This develops students' confidence, reminding them of their achievements and providing them with specific 'career stories' to demonstrate critical skills during their job search, interviews, and conversations with potential employers. Students are prompted to consider their 'About Me'—including their motivations to

Table 3 Overview of the elements of the portfolio assessment in the ICT WIL Placement course, including alignment to learning outcomes

Included in your portfolio		Word Guide	Weighting	Learning Objective
1	Critical analysis of WIL placement	600–800 words	25%	1, 2, 3
2	Evidence of technical and transferable skills/ knowledge	600–800 words	25%	1,2
3	Outline of placement deliverable and evidence of decision making	500–600 words	20%	1, 2, 3
4	Evidence of your awareness of ethical conduct in your industry	250–400 words	10%	3
5	About Me statement including area of study and career aspirations	200–300 words	5%	4
6	Holistic experience artefact (extra-curricular experience), linked to key skills	200–300 words	5%	2
7	Overall presentation of your portfolio including ease of navigation, use of appropriate visual elements, professional language and free from errors	N/A	10%	4

study in ICT and their career aspirations- and to include a 'holistic experience' artefact reminding students of the diverse life experiences they can draw on to prove their capabilities.

5.3 Challenges

In each of the artefacts included in the 3822ICT portfolio, students are encouraged to *critically evaluate* their experiences during their placement, to describe what they did and why; to explain decisions they made and how effective these decisions were, to question and assess their deeply held assumptions about their knowledge and the way they perceive events; and to identify an alternative outcome if different decisions had been made or different actions undertaken. Initially, students did poorly on this part of the task. While they provided a reasonable reflection about their experiences, they did not shift to 'critical reflection' to include areas of strength (what they did well) and areas for improvement. This is an important transferable skill for students to ensure they are independent learners, allowing them to transfer their existing knowledge and skills and critically evaluate the outcomes of their actions. To address this issue, additional resources (workshops and written resources in PebblePad) were added to help students learn how to reflect critically. This has led to improvements in the quality of student work.

A unique element of the portfolio assessment in 3822ICT is the inclusion of a specific reflection that evidences 'technical and transferable skills', with significant emphasis on the actual transference of skills via prompts for the students as follows:

- What technical skills/knowledge have you developed, enhanced or applied through your placement?
- Is this an application of an existing skill/knowledge set to a different context? If so, explain the difference in context and what you have learned by doing this.
- What specifically did you do that allowed you to further develop your skills/knowledge?
- What prior knowledge did you draw/build on to help you apply your skills/knowledge during your placement?

Following feedback from students that they felt uncomfortable, including 'negatives' or 'criticisms' of themselves in the portfolio, greater emphasis was placed on the learning process, which often includes 'trial and error' before success. This empowered students to turn a potential negative into solid evidence of their adaptability and capacity for life-long learning via critical self-evaluation, including identifying instances of transference.

5.4 Outcomes and Evaluation

Substantial evidence is that writing reflections complements students' placement and learning experience. Reflective practice empowers students to identify their strengths and weaknesses in a real-life work environment. The recommended structure of reflective writing for 'showcasing their experiences for others' using a known reflective framework ensures students capture and articulate specific details about what they have learnt from their experiences, including an overt focus on critical transferable skills required for workplace success.

One strong example is drawn from a student's portfolio:

> …it was not a totally smooth process and added some unnecessary complexities. To avoid this, I should have talked to more of the team … so that I could gain a better understanding of what was happening and how I could fit my project into [theirs]. This taught me a lot about the importance of talking to colleagues and gaining a better understanding of the situation before rushing in and making decisions that could cause difficulties. In the future, I plan to be much more open with colleagues and work on my general discussion skills so that those conversations can happen with more ease and frequency. (Trimester 2, 2023)

Another example evidences deep awareness about the impact of decision-making:

> The journey was challenging and not without struggle. Reflecting on my actions, I realized there were areas where I could have achieved a better outcome if I had taken different steps. (Portfolio reflection, Trimester 2, 2023)

Many students reflected on the importance of "continuous learning and personal development" (Portfolio Reflection, Trimester 2, 2023) within their portfolio, while others captured an overall view of the impact of the portfolio assessment, specifically the reflective writing component:

> Reflective writing and learning have given me a new understanding of my current state of self (SEC, 7822ICT, Trimester 1 2022)
>
> I felt the focus on continuous learning and ongoing personal development in the workforce was perfect. It brought me back to the reason I do what I do and made me appreciate more the daily decision-making processes...The portfolio activity was a great practical exercise...-it helped me to identify a different approach to presenting information. (SEC, 7822ICT, Trimester 1 2022)

Although there are many strong insights from students within their reflections about what they have learned and why it is important, there are some students who still struggle to identify the relevance of the reflective writing process:

> The amount of reflections for the portfolio felt too much. The usefulness of the reflections as a portfolio is questionable as it is not designed to target future employers. (3822ICT, SEC, T2 2023)

This comment is balanced by other students who said, "The portfolio is a fantastic piece of work that can be used for future employers"(3822ICT, SEC, T1 2022) and "The focus on improving employability is valuable for students" (3822ICT, SEC T2, 2023). However, it does indicate an ongoing need to educate students about the value of reflective practice for their career development from the beginning of their degree, throughout their studies and into their professional roles. One key goal is to shift the focus of portfolio development, from the student perspective, from the sphere of 'documentation of achievement' to emphasize 'documentation of learning' (Barrett, 2010). This shift will ensure students understand the benefits of *creating* their portfolio as being as important (if not more important) than an unknown future opportunity to share their portfolio with an invested employer.

6 STEM Professionals—Co-Curricular Professional Development Program

6.1 Course and Participant Overview

STEM Professionals is a co-curricular career readiness program introduced in the Sciences Group (Faculty) in 2014 as Griffith Sciences PLUS (Professional Learning for University students). The program was renamed in late 2023 to better capture the alignment with all STEM disciplines within the Group and to emphasize the relationship between the concepts introduced during workshops and reflective activities to the students' developing professional identity. STEM Professionals is available to all students within the Sciences Group, with the ultimate achievement—"Gold STEM Professionals+ "—directly linked to each student's discipline, as depicted in Fig. 5. Students can commence their participation in STEM+ at any stage of their degree program. Workshop content and activities are designed to allow students to align concepts and intended actions to their current professional readiness level, previous career experiences, and aspirational career goals.

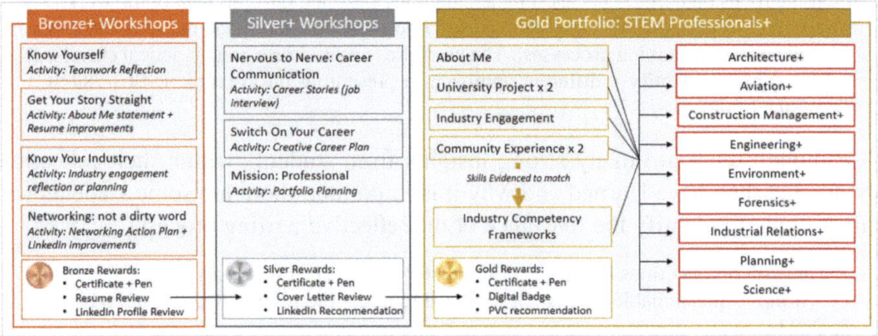

Fig. 5 STEM Professionals+ Program Structure and Discipline-based awards

6.2 Implementation of Portfolio Assessment

Rationale and competencies: The development of the student's portfolio is integral to the objectives of STEM Professionals and is designed to empower students to reframe events and experiences from their lives to recognize and ultimately articulate their unique skills, capabilities and knowledge (Brent, 2019). The process aligns with Barrett's (2010) notion of the "two faces of ePortfolio", which emphasizes the relevance of both the learning and showcase portfolios. The notion of 'transference' is key here as many students struggle to connect 'life' experiences with capabilities they will need for success, firstly at university and, ultimately, in the workplace. Without an opportunity to reflect, they can dismiss or devalue their existing skills as irrelevant to their future career. The activities in STEM Professionals aim to address this through purposeful reflective practice connected to workshop content in the key themes of career management, establishing professional networks, developing professional identity, creating a solid career toolkit, and evidencing skills to meet industry expectations.

Connection to Industry Competency Frameworks: The STEM Professionals workshops are delivered simultaneously to students in all STEM disciplines within the Sciences Group. Content examples are derived from relevant STEM disciplines; however, throughout the workshop activities, the onus is on the student to connect to specific discipline interests. This overtly shifts to greater emphasis on discipline alignment in the final STEM+ workshop "Mission: Professional" when students are introduced (or reintroduced) to competency frameworks attached to STEM disciplines—for example, the Engineering Stage 1 Competencies, Planning Institute of Australia Competencies, the Australian Computer Society Core Body of Knowledge, Science Threshold Learning Outcomes.

Structure of the Portfolio: To achieve Gold+ , students must align elements in their portfolio (artefacts: reflections and visual elements) with the competency framework appropriate for their degree. Student reflective practice within STEM Professionals is scaffolded throughout the program to enable them to achieve this. After each workshop, students complete an activity relevant to the workshop content. The screenshot in Fig. 6 shows the functional use of PebblePad to provide students with weekly, meaningful feedback and help both students and educators track students' progress through the program.

At the Gold+ level, the notion of transference and an acknowledgement of skills developed in unstructured learning environments is included via the requirement for students to include artefacts capturing at least one example of a 'community' experience in addition to evidence from within the curriculum and evidence of at least one interaction with industry (shown in Fig. 5). This encourages students to assign meaning to their individual experiences in a purposeful way. Notably, few restrictions exist on what constitutes a 'community' experience. The expectation is that if students can explain why an experience was essential to them and if they can draw from that experience to provide tangible evidence of skills developed, it is appropriate for inclusion in the portfolio.

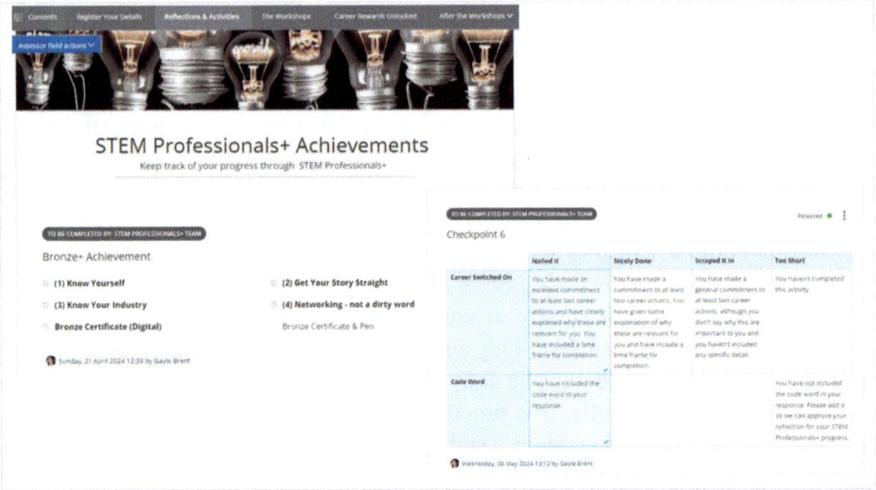

Fig. 6 Use of PebblePad functionality to track student progress and ensure timely feedback is provided after each workshop

6.3 Challenges

The current STEM Professionals structure was established in 2019, shortly after Griffith University launched access to PebblePad. At that time, the program was reframed from a points-based, self-guided series of modules to the workshop series delivered in the stages of Bronze+ , Silver+ and Gold+ (or Bronze PLUS, Silver PLUS and Gold PLUS at it was then) as depicted in Fig. 5. One significant challenge was inspiring students to achieve the higher levels of the program. The number of students earning Silver+ in 2020 was substantial at 50, with 12 students earning Bronze+ . A concern was that only 7 (or 12%) of eligible Silver+ students went on to achieve Gold+ , as depicted in Fig. 7.

Given the exceptional benefits for students in portfolio development, it was imperative to restructure the program to inspire more students to continue their Gold+ achievement. This was first achieved via the introduction of regular feedback on student reflections following workshops, using the functionality of PebblePad. Additional measures included overt messaging of the alignment between Gold+ and the work undertaken in the WIL, practicum, and placement courses profiled in the case studies in this chapter, which were introduced to students in the associated portfolio activity in their first year. There was a significant increase in completion rates for Gold+ , from 12% of eligible students in 2020 to 40% of eligible students in 2021 (Fig. 7). Despite increasing numbers of students engaged in the program (69, 101, 117, 154 in 2020, 2021, 2022 and 2023, respectively), the lower percentage of Gold+ completions in 2022 (27% of eligible students) highlighted the need for change. In 2023, the "after workshop" activities were reinvigorated to ensure each activity played a dual role: capturing student ideas and responses to the workshops

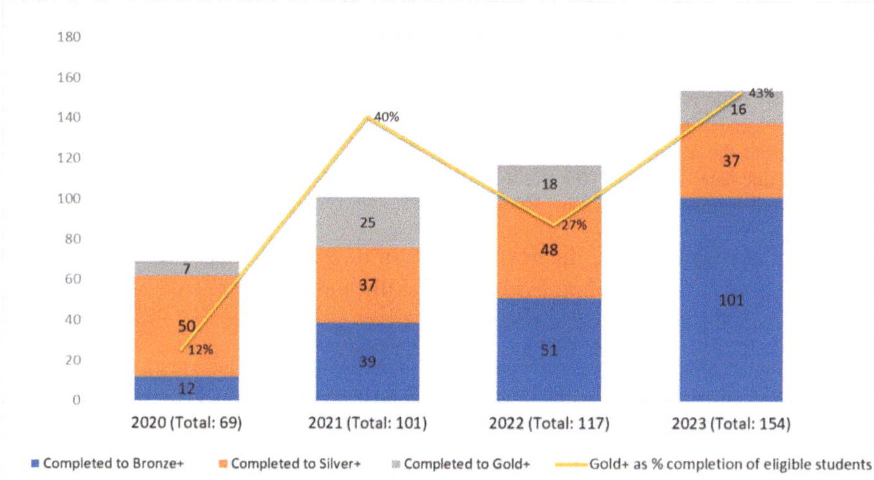

Fig. 7 STEM Professionals Achievements 2020–2023, including percentage Gold + completions of eligible students

and allowing them to produce a usable artefact they could later include in their portfolio. This appears to have been a successful strategy. Percentage completion of Gold+ improved from 27% in 2022 to 43% in 2023, with early indicators that this trend will continue in 2024.

6.4 *Outcomes and Evaluation*

Students in the Griffith University STEM Professionals program provide consistently positive feedback about the impact of the workshop and reflective activities on their confidence and career preparedness. With respect to evidence of the impact of the portfolio component on students, several key themes emerge based on student feedback.

When undertaken in their early university years, the simultaneous development of the student's portfolio, alongside 'regular' curricular learning and assessment, allows them to gain greater insight into their learning and to attach additional value to their university work:

> What stood out straight away for me was the idea of the creative job search. Being reminded that your career development is happening every day was helpful because it **gives intrinsic value to every bit of university work I do.** (First-year ICT Student, 2022)

Throughout their study, creating their portfolio enables students to perceive connections between their university work and assessments, from course to course, year to year. It empowers them to cohesively draw from these examples of their work to demonstrate their capability. It provides an opportunity for them to evidence their skills in the context of professional expectations (accountability):

> The showcasing portfolio has a primary focus on accountability…This interested me as I [have not looked] back at my previous assignments as products but rather [as] tasks. I would assume that the assignments are only helpful for the assessment task and hold no value after the due date. **Whereas [the portfolio] approach allows previous work to hold value.** (Third-year Engineering Student, 2023)

As they near graduation, the development of their portfolio provides an opportunity for students to collate their evidence of skills, knowledge and capability to draw from these during their active job search:

> All [of] the PLUS sessions and activities…got me thinking about interviews…and helped with [my] Cadetship interview. **Having all the experiences from my portfolio was a huge help** and no doubt helped me get to this point. (Final year Aviation student, 2021)

Portfolio development also inspires students to perceive their achievements through an individual lens that aligns with their emerging professional identity:

> **This portfolio illustrates my experiences as a collection of unique skills**, capabilities, knowledge, personality, and aspirations **for my future career.** (Final year Environmental Science student, 2021)

Including the portfolio in the STEM Professionals program drives students' consistent engagement with the workshops and workshop activities. The tangible outcomes of developing their portfolio and the associated impacts on their career confidence and preparedness to transition from university to the workplace are clear to students and act as intrinsic motivators for them to achieve Gold+ . While not all students reach this level (this is a voluntary program, after all), all students who engage in STEM Professionals can still gain significant career-readiness benefits based on the staged approach to portfolio development via the workshop activities, complete with meaningful feedback.

7 Discussion

The complex objectives of higher education institutions are to facilitate learning for diverse learners whilst simultaneously recognizing the broad gamut of experiences they have had prior to and concurrent with their participation in higher education. This establishes the catalyst for change, as purported by Berger and Wild (2017). They assert that the shift to appropriate methods for assessment of 'wicked competencies' (or transferable skills) developed in authentic contexts via innovative learning methods can only occur if there is first a shared understanding (amongst all stakeholders) that the inadequacies of current assessment practice are simply 'not good enough' (p. 431).

By including reflective tasks such as portfolios, student perceptions of their university assessments as a set of siloed and disconnected assignments, quizzes, exams, and lab tests could instead be the foundation of a transformative collection that allows students to individually evidence their skills, knowledge and capabilities, as they develop over time, in context of their discipline. Ferns and Comfort (2014) and Carter (2021) support this view of portfolio with integrated reflective practice, highlighting portfolios in the curriculum as a potential solution to the 'tension' between practising transferable skills and evidencing them to improve employability.

We argue, and the case studies in this chapter provide evidence, that embedded portfolio assessments empower students to develop greater confidence and transfer their skills to new contexts. If students can practice the 'transfer' of skills and knowledge from life experience to university, from one course to another, one assignment to the next, from structured university learning to unstructured workplace learning during WIL, it follows that, as graduates, they will ultimately have the capacity to actively transfer their skills to new contexts in the workplace in a graduate role and throughout their career. We suggest that the 'ability to transfer' is a transferable skill in and of itself. As educators, we should seek practical opportunities to facilitate student learning of this skill just as we would seek opportunities to teach any other transferable skill. To achieve this, students "receive the appropriate pedagogical support to meaningfully reflect on and articulate the transferable skills they have developed" (Oguro & Mueller, 2020).

Portfolio assessment tasks are not without their challenges, as they may be new to students and teaching teams. It is crucial that educators consider how to communicate the value of portfolios to students, and we suggest a 'what is in it for me' type explanation may be a helpful approach. Having students link their reflections to the relevant industry competencies can also help students see links between the task and the expectations of their future professions (Howell et al., 2023). We also recommend that students be exposed to reflective tasks throughout their degree rather than leaving the reflective portfolio to the final year.

8 Conclusion

We have established that a) the development and evaluation of competencies and capabilities developed within WIL (and in broader life experiences) can be challenging to quantify and assess; b) traditional methods of assessment are often inadequate for ill-defined and complex competencies that include employability and c) that holistic approaches to assessment that empower students to connect their experiences and articulate a narrative of their employability is crucial to align with employer expectations. This leads to further observation regarding the apparent skills gap, which arguably exists, not because students do not have the skills but because they either do not recognize their skills and/or they cannot confidently articulate them. The nature of WIL allows students to apply and develop their discipline skills, particularly their transferable skills, in the context of their discipline. It is, therefore, a mechanism to close the skills gap. We contend that innovative portfolio-based assessments are a powerful, potential solution to overcome 'skills gaps', either actual or perceived. This is achieved via the purposeful use of reflective practice throughout the placement, in combination with the articulation of skills and capabilities within the portfolio to enable students to 'match' and 'transfer' their existing skills and knowledge (including crucial transferable skills such as communication, critical thinking, problem-solving and decision-making) to novel situations in the workplace.

References

Akkerman, S. F., & Bakker, A. (2011). Boundary crossing and boundary objects. *Review of Educational Research, 81*(2), 132–169. https://doi.org/10.3102/0034654311404435

Akkerman, S. F., & Van Eijck, M. (2013). Re-Theorizing the student dialogically across and between boundaries of multiple communities. *British Educational Research Journal, 39*(1), 60–72.

Baldwin, C., & Rosier, J. (2017). Growing future planners: A framework for integrating experiential learning into tertiary planning programs. *J. Planning Education & Research, 37*(1), 43–55.

Barrett, H. C. (2010). Balancing the two faces of ePortfolios. *Educação, Formação e Tecnologias, 3*(01), 06–14.

Baston, T. (2010, July 28). *A profoundly disruptive technology*. Campus Technology. https://campustechnology.com/Articles/2010/07/28/A-Profoundly-Disruptive-Technology.aspx

Bell, J. (2024). *What stage of leadership competence? Results-wise consultancy*. https://www.resultswise.com.au/what-stage-of-leadership-competence/

Bennett, D., Rowley, J., Dunbar-Hall, P., Hitchcock, M., & Blom, D. (2016). Electronic portfolios and learner identity: An ePortfolio case study in music and writing. *Journal of Further and Higher Education, 40*(1), 107–124. https://doi.org/10.1080/0309877X.2014.895306

Berger, D., & Wild, C. (2017). Enhancing student performance and employability through the use of authentic assessment techniques in extra and co-curricular activities (ECCAs). *The Law Teacher, 51*(4), 428–439.

Blaquiere, G. de, Nolan, J. E., & Wray, K. (2019). Joining up the dots: telling the story of employability. How can students in higher education be supported to better understand and articulate their employability? *Journal of Teaching and Learning for Graduate Employability, 10*(2), 15–35.

Bransford, J., & National Research Council (U.S.). Committee on Developments in the Science of Learning. (2000). *How people learn: Brain, mind, experience, and school (Expanded).* National Academy Press. February 14, 2024.

Brent, G. (2019). *Creating order from (potential) chaos: Embedding employability with the Griffith Sciences PLUS program* (pp. 99–119). Springer. https://doi.org/10.1007/978-981-13-6982-7_6

Bridgstock, R. (2009). The graduate attributes we have overlooked: Enhancing graduate employability through career management skills. *Higher Education Research & Development, 28*(1), 31–44.

Bridgstock, R., & Tippett, N. (2019). *Higher education and the future of graduate employability: A connectedness learning approach.* Edward Elgar Publisher.

Bridgstock, R., Grant-Iramu, M., & McAlpine, A. (2019). Integrating career development learning into the curriculum: Collaboration with the careers service for employability. *Journal of Teaching and Learning for Graduate Employability, 10*(1), 56–72.

Brown, P., & Hesketh, A. (2005). The mismanagement of talent: Employability and jobs in the knowledge economy. *Industrial & Labor Relations Review,* 50. https://doi.org/10.2189/asqu.2005.50.2.306

Carter, S. (2021). EPortfolios as a platform for evidencing employability and building professional identity: A literature review. *International Journal of Work-Integrated Learning, 22*(4), 463.

Choate, J., Demaria, M., Etheve, M., Cran, S., & Carroll, D. (2019). A professional development program with an assessed ePortfolio: A practical solution for engaging undergraduates with their career development in large student cohorts. *Journal of Teaching and Learning for Graduate Employability, 10*(2), 86–103.

Clanchy, J., & Ballard, B. (1995). Generic skills in the context of higher education. *Higher Education Research and Development, 14*(2), 155–166.

Cushing, D. F., Pennings, M., Willox, D., Gomez, R., Dyson, C., & Coombs, C. (2019). Measuring intangible outcomes can be problematic: The challenge of assessing learning during international short-term study experiences. *Active Learning in Higher Education, 20*(3), 203–217.

Dewey, J. (1998). *The essential Dewey, Volume 1: pragmatism, education, democracy* (Vol. 1). Indiana University Press.

Faulkner, M., Mahfuzul Aziz, S., Waye, V., & Smith, E. (2013). Exploring ways that ePortfolios can support the progressive development of graduate qualities and professional competencies. *Higher Education Research & Development, 32*(6), 871–887. https://doi.org/10.1080/07294360.2013.806437

Ferns, S., & Comfort, J. (2014). Eportfolios as Evidence of Standards and Outcomes in Work-Integrated Learning. *Asia-Pacific Journal of Cooperative Education, 15*(3), 269–280.

Griffith University. (2024). *Griffith employability framework*. https://www.griffith.edu.au/careers-employment/staff/griffith-employability-framework

Guàrdia, L., Maina, M. F., & Mancini, F. (2021). Increasing the visibility of graduate students' employability skills: An ePortfolio solution addressing the skills gap. In *Cases on active blended learning in higher education* (pp. 253–275). IGI Global.

Harvey, L. (2000). New realities: The relationship between higher education and employment. *Tertiary Education and Management, 6*(1), 3–17. https://doi.org/10.1023/A:1009685205201

Howell, S., Hall, W., & Geelan, D. (2023). Exploring the perspectives of engineering undergraduates on employability and employability building activities. *Higher Education, Skills and Work-Based Learning, 13*(1), 161–178. https://doi.org/10.1108/HESWBL-05-2022-0111

Jackson, D. (2014). Testing a model of undergraduate competence in employability skills and its implications for stakeholders. *Journal of Education and Work, 27*(2), 220–242. https://doi.org/10.1080/13639080.2012.718750

Jackson, D., & Bridgstock, R. (2021). What actually works to enhance graduate employability? The relative value of curricular, co-curricular, and extracurricular learning and paid work. *Higher Education, 81*(4), 723–739. https://doi.org/10.1007/s10734-020-00570-x

Nadelson, L. S., McGuire, S. P., Davis, K. A., Farid, A., Hardy, K. K., Hsu, Y. C., & Wang, S. (2017). Am I a STEM professional? Documenting STEM student professional identity development. *Studies in Higher Education, 42*(4), 701–720.

Oguro, S., & Mueller, B. (2020). Learning abroad and graduate employability: Challenges articulating international learning outcomes. In E. Heinrich & R. Bourke (Eds.), *Research and development in higher education: Next generation, higher education: Challenges, changes and opportunities*, 42 (pp. 85—93).

Planning Institute of Australia. (PIA). (2020). *Policy for the accreditation of Australian planning qualifications*. https://www.planning.org.au/documents/item/48

Reid, A., Richards, A., & Willox, D. (2021). Connecting experiences to employability through a meaning-making approach to learning. *Journal of Teaching and Learning for Graduate Employability, 12*(2), 99–113.

Reis, K., Brent, G., & Martin, S. (2023). Collaborative local food resilience planning in Logan, Australia. In M. Heim & J. Mittal (Eds.), *Routledge Handbook of University-Community Partnerships in Planning Education* (pp. 305–327). Routledge.

Perrone, L., & Vickers, M. H. (2003). Life after graduation as a "very uncomfortable world": An Australian case study. *Education + Training, 45*(2), 69–78. https://doi.org/10.1108/00400910310464044

Prinsley, R., & Baranyai, K. (2015, August). STEM-trained and job-ready. *Office of the Chief Scientist Occasional Paper Series* (12). Retrieved from https://www.chiefscientist.gov.au/wp-content/uploads/OPS12-WIL_web.pdf

Sarkar, M., Overton, T., Thompson, C., & Rayner, G. (2016). Graduate employability: Views of recent science graduates and employers. *International Journal of Innovation in Science and Mathematics Education, 24*(3), 31–48.

Schon, D. (1983). *The reflective practitioner: How professionals think in action.* Taylor and Francis.

Simatele, M. (2015). Enhancing the portability of employability skills using e-portfolios. *Journal of Further and Higher Education, 39*(6), 862–874. https://doi.org/10.1080/0309877X.2014.953461

Spanjaard, D., Hall, T., & Stegemann, N. (2018). Experiential learning: Helping students to become 'career-ready'. *Australasian Marketing Journal (AMJ), 26*(2), 163–171.

Taylor, E., & Hurley, J. (2016). "Not a lot of people read the stuff," Australian Urban Research in Planning Practice, *Urban Policy and Research, 34*(2), 116–131.

Wenger, E. (1998). *Communities of practice: Learning, meaning, and identity.* Cambridge University Press.

Yorke, M., & Knight, P. T. (2006). *Embedding employability into the curriculum* (Vol. 3). Higher Education Academy.

Chapter 8
Authentic Learning and Student's Employability: Are the Work Integrating Learning Courses Really Working?

Karine Dupre, Sebastian Smith, Kailani Hubbard, Ting Ren, and Julie Crough

Abstract Student employability is a significant focus for universities, industry and national governments. Work-integrated Learning (WIL) represents an authentic educational approach that bridges theoretical knowledge with practical experience. It has been widely used as an effective and efficient measure to enhance graduates' employability. Nonetheless, empirical evidence shows that few students enrol in this type of course when it is not part of the required curriculum. This study investigated 29 WIL courses at Griffith University (Australia) to gain deeper insights into this issue. The objective was to identify trends and gaps across disciplines, understand WIL's structural elements, such as WIL types, learning outcomes, and graduate attributes, and gather feedback from key stakeholders, including course convenors, industry partners, and students. The findings demonstrate that while there is no doubt about the positive impact of WIL on employability, there is a pressing need to foster a genuine WIL culture and refine certain aspects of the current WIL framework. These steps are crucial to facilitate, enhance, and promote more meaningful WIL experiences in the future.

1 Introduction

Work-integrated Learning (WIL) interweaves practical work experience with classroom learning, which bridges theory and practice for graduates or students (Jackson, 2017; Patrick et al., 2008). Originating in Australia (Miller et al., 2022), WIL has been widely used by higher educational institutions worldwide. Ferns and Lilly (2015, p. 118) underscore how WIL is recognized as "*a key strategy for universities to address the demands of stakeholders such as industry, community, students and government*". As an authentic learning approach that embeds meaningful and 'real

K. Dupre (✉) · S. Smith · K. Hubbard · T. Ren · J. Crough
Griffith University, Gold Coast, QLD, Australia
e-mail: k.dupre@griffith.edu.au

I. Gratchev et al. (eds.), *Transformative Approaches to Career-Ready Education*,
https://doi.org/10.1007/978-981-96-7018-5_8

world' industry placements into an academic program, WIL is essential in transitioning from student to professional practitioner. WIL practices have achieved better employment outcomes in Australia (Ferns et al., 2014). As an authentic learning approach, WIL encourages students to reflect on and connect with mentors through their experience, thus broadening their career horizons (Prinsley & Baranyai, 2015). While industry placements are a requirement for Engineering programs in other science, technology, engineering and mathematics (STEM) disciplines, they are often not compulsory. They are provided ad hoc or on a small scale (Prinsley & Baranyai, 2015).

The increasing expectation of implementing WIL is to enhance graduates' employability, generally defined as a set of skills and personal attributes that will prepare graduates for the workplace and make successful career progression (Jackson & Collings, 2018). The set of employability skills is generally categorized into hard and soft skills. Hard or core skills are specific and necessary technical skills and knowledge typically learned through formal education or training (Bridgestock, 2009). With exposure to real-world work environments and opportunities to apply theory to practice, WIL can help students develop complex skills essential for success in the workforce. Depending on the chosen profession, examples of hard skills in the sciences include proficiency in programming languages, database management, conducting experiments, and architectural or graphic design. These skills are often specific to a particular job or industry and can be measured more objectively.

On the other hand, soft skills or generic skills are interpersonal or people skills that are often harder to measure or quantify. Soft skills include effective communication, teamwork, leadership, adaptability, problem-solving, and critical thinking. These skills are often transferable across different roles and industries and are highly valued by 21st-century employers (Edziwa & Bilgnaut, 2022; Ferns & Lilly, 2015). Similarly, personal attributes are difficult to measure or quantify. Edziwa and Blignaut (2022) define personal attributes as self-awareness, emotional intelligence, resilience, adaptability and a positive attitude, which can influence how a person approaches work and interacts with others.

Compared to hard skills, the recent emphasis in education—especially WIL—has shifted to developing soft skills (Jackson & Wilton, 2016). WIL has been confirmed to positively contribute to developing those skills, particularly interpersonal skills, communication, and teamwork, for those participating in the program (Falconer & Pettigrew, 2003). Furthermore, WIL can facilitate the development of students' pre-professional identity through networking, observing, questioning and interacting with professionals in the workplace (Jackson, 2017). Specifically, Jackson (2017) affirms that WIL can help students develop a more comprehensive understanding of their chosen profession, including its responsibilities, expectations and standards, attitudes, beliefs, and ethical values. WIL also positively influences students' professional network development (English et al., 2021). Altogether, WIL helps students become more effective and ethical professionals and better understand the impact of their work in the broader community. Similarly, Jackson and Wilton (2016) ascertain that students participating in WIL programs have better performance and ratings for career self-management competencies, which is an essential soft skill, compared with those who do not participate.

However, the success of WIL programs depends heavily on the cooperation and partnership between universities and industry or organizations. Referring to WIL as cooperative education, Kessels and Kwakman (2007) emphasize the importance of building sustainable relationships between universities and industry partners and working together to create learning opportunities and facilitate students to link theoretical knowledge with practice. Thune and Storen (2015) have shown that cooperative partnerships between universities and industry can positively influence the graduate learning experience and position graduates in a better place in the labour market. To ensure that WIL programs are effective, universities and industry partners need to clearly understand the WIL process and their respective roles and responsibilities (Gibson et al., 2016). This includes understanding the program's learning outcomes, the expectations for students and industry partners, and the assessment and evaluation processes. In addition, both parties should be committed to maintaining open communication and collaboration throughout the WIL process (Gibson et al., 2016). This includes regular communication between students, university staff, and industry partners and ongoing feedback and evaluation to ensure the program meets all stakeholders' needs.

However, while some studies have discussed the assessment process or proposed structure and plan to examine the outcomes of WIL programs in their chosen fields (Ferns & Lilly, 2015; Gaskell & Beaton, 2010; Scanlan, 2008; Vaughan & Moore, 2016), there is still a lack of comprehensive and consistent evaluation across different programs and contexts. In other words, the reviewed literature shows that the majority of WIL programs rely on a limited set of assessment methods, such as supervisor evaluations (Fleming & Pretti, 2019; Fynn et al., 2019), reflective essays or students' self-reflection (Benati et al., 2021; Jackson, 2017) or questionnaires (Alm et al., 2022; Falconer & Pettigrew, 2003; Jackson & Collings, 2018; Marshall, 2012; Thune & Storen, 2015). This is surprising, considering limited assessments or evaluation methods cannot fully capture the breadth of skills and knowledge that students develop during their WIL experiences. This can make it challenging to identify areas for improvement and ensure that WIL programs meet all stakeholders' needs. Furthermore, consistent evaluation is necessary to identify best practices for designing and implementing effective WIL programs.

Therefore, this study aims to explore the effectiveness, inclusivity, success and efficiency of WIL through a case study approach within Griffith University and reveal existing main trends and gaps across several disciplines, understand the WIL structural elements (such as types, learning outcomes, graduate attributes), and feedback from key stakeholders including course convenors, industry partners, and students. Overall, it provides an overview of the current trends and gaps anchored in theoretical discourses and practices while questioning the current WIL framework. Structurally, this chapter is divided into three sections: the first section presents the methodology implemented in this research; the second section reports the results; and the third section, the discussion, is based on the triangulation of the main findings with insights from the existing scholarship.

2 Method

2.1 Overview

This research used a mixed-method comparative case study approach, investigating 29 WIL courses at Griffith University, all under the Sciences Group except for eight within Queensland College of Arts. These courses were selected for convenience (the research team works at this University and is all affiliated with the Sciences Group) and to address a disciplinary gap identified through a preliminary systematic literature review of WIL practices. Among the 37 articles reviewed (see Table 1), health represented the field with the most conducted research (10/37), followed by business (6/10), while Art and Sciences were comparatively underrepresented. This literature review was also used to identify and define the existing WIL types, like the typology proposed by Jackson et al. (2022). This provided a valuable reference for comparison. The five WIL types are:

(1) **internships/placements/practicum** (spending time completing tasks/projects in the workplace)
(2) **projects/consultancies** (working on real-live projects/strategies/briefs provided by industry/community partners)
(3) **service learning** (working on community-focused projects/tasks)
(4) **incubator/start-ups** (working on distinct phases of a business start-up with industry support)
(5) **simulation**

The mixed method study specifically investigated two main aspects. The first was the contextual and pedagogical setting of WIL offering at [name withhold] University based on a desktop study. It included understanding the institutional framework, the characteristics of the WIL courses (discipline, type) and the potential correlations between the WIL type, learning outcomes (LOs) and graduate attributes (GAs). A face-to-face workshop with educators complemented this desktop study. The second key aspect investigated concerned another desktop study focusing on the student's background in one discipline, specifically (architecture) to evaluate EDI's implementation. The following paragraphs define WIL type, LOs and GAs and present each protocol in detail.

2.2 WIL Type, LOs and GAs

Learning outcomes are the central organizing feature of student learning at a program or course level. At a course level, learning outcomes and their aligned assessment items are integral to the knowledge and skills, as well as their application, that a student should be able to exhibit and demonstrate upon successful course completion. Learning outcomes integrate depth, breadth, complexity, types of disciplinary

Table 1 WIL articles for literature review

Number	Article/Paper	Author/s	Year	Discipline
1	Newly graduated nurses' work-integrated Learning: A qualitative study from an educational and occupational perspective	Andersson, A., Hallgren U., Graneheim, M. Skyvell, N	2022	Health (nursing and midwifery)
2	Innovation in Australian nursing student work-integrated learning during a pandemic	Sue Dean, Darrelle Ahchay, Felicity Walker, Colleen Ryan	2022	Health (nursing and midwifery)
3	Exploring the contribution of clinical placement to student learning: A sequential mixed methods study	Tracey Moroney, Marie Gerdtz, Naomi Brockenshire, Phil Maude, Jennifer Weller-Newton, Deborah Hatcher, Luke Molloy, Moira Williamson, Robyn Woodward-Kron, Elizabeth Molloy	2022	Health
4	Work-integrated Learning as a pedagogical tool to Integrate Theory and Practice in Nursing Education—An Integrative Literature Review	Ina Berndtsson, Elisabeth Dahlborg, Sandra Pennbrant	2020	Health (nursing and midwifery)
5	The mini Clinical Evaluation Exercise (mini-CEX) in a pre-registration osteopathy program: Exploring Aspects of its Validity	Brett Vaughan, Keri Moore	(2016)	Health (Osteopathy)
6	Learning to learn through University accredited work-based Learning: A threshold conception	Charlotte Ramage	(2014)	Health (nursing and midwifery)
7	Developing midwifery practice through work-based Learning: An exploratory study	Jayne E. Marshall	(2012)	Health (midwifery)
8	Inter-professional work-based Learning within an MSc in Advanced Practice: Lessons from one UK Higher Education programme	Lynne Gaskell, Susan BeatoN	(2010)	Health
9	A theoretical model for supporting bounded learning communities focussed on promoting clinical reasoning in health students undertaking field-based education	JN Scanlan	2008	Health

(continued)

Table 1 (continued)

Number	Article/Paper	Author/s	Year	Discipline
10	Developing the health care workforce- a comparison of two work-based learning models			Health
11	Norbert Dentressangle delivers managers of tomorrow	Norbert Dentressangle	2010	Business
12	Applying theory in practice: views of graduating business students	Kelly Benati, Sophie Lindsay, Juan Fischer	(2021)	Business
13	Developing professional networks: the missing link to graduate employability	Peter English, Margarietha Johanna de Villiers Scheepers, David Fleischman, Jacqueline Burgess, Gail Crimmins	2021	Business
14	Developing pre-professional identity in undergraduates through work-integrated Learning	Denise Jackson	(2017)	Business
15	Accelerating supply chain management learning: Identifying enablers from a university-industry collaboration	Trish Gibson, Donald Kerr, Ron Fisher	(2016)	Business
16	Developing Career Management Competencies among Undergraduates and the Role of work integrated Learning	Denise Jackson, Nicholas Wilton	(2016)	Business
17	Students learning sustainability—implicit, explicit or non-existent: a case study approach on students' key competencies addressing the SDGs in HEI program	Karin Alm, Thomas H. Beery, David Eiblmeier, Tarek Fahmy	2022	Multiple
18	Design principles for electronic work integrated Learning (eWIL)	Charmaine Glavas, Lisa Schuster	2020	Multiple
19	The influence of Work-Integrated Learning and paid work during studies on graduate employment and underemployment	Denise Jackson, David Collings	(2018)	Multiple
20	Driving institutional engagement in WIL: Enhancing graduate employability	Sonia Ferns, Linda Lilly	(2015)	Multiple

(continued)

Table 1 (continued)

Number	Article/Paper	Author/s	Year	Discipline
21	Study and Labour Market Effects of graduate students' Interaction with work organizations during education	Taran Thune, Liv Anne Støren	2015	Multiple
22	Earnings and employment outcomes for male and female postsecondary graduates of coop and non-coop programmes	David Walters, David Zarifa	(2008)	Multiple
23	Curriculum innovation in undergraduate accounting degree programmes through "virtual internships"	Leopold Bayerlein	(2015)	Accounting
24	Graduate employability skills: the voice of agricultural technical vocational education and training (ATVET) students in Zimbabwe	X. Edziwa, S. Blignaut	2022	Agriculture
25	Delivering distance education for modern government: The F4Gov Programme	Jon Talbot	(2007)	Civil Service
26	The work-based learning experience of trainee teachers in English further education colleges	Kevin Orr, Robin Simmons	(2011)	Education
27	New models for improving teaching and training in engineering and technology transfer in South Africa	L. W. Snyman, D. Kruger	(2020)	Engineering
28	Developing added value skills within an academic programme through work-based Learning	Stewart Falconer, Malcolm Pettigrew	2003	Finance & Economy
29	A study on the skills of a Master of Arts in Research Psychology graduates from the employers' perspective	A. W. R. Fynn, S. N. Dladla, H. Erasmus	2019	Psychology
30	An assessment of work-integrated Learning for public relations in an open distance learning context	Christelle Swart	2014	Public Relations
31	"Casing the joint" Explorations by the insider researcher preparing for work-based projects	Barbara Workman & Costley, C	2007	Research

(continued)

Table 1 (continued)

Number	Article/Paper	Author/s	Year	Discipline
32	Problems experienced by students on field placement: using research findings to inform curriculum design and content	Jane Maidment	(2010)	Social Work
33	The impact of work-integrated learning students on workplace dynamics	Jenny Fleming, T. Judene Pretti	(2019)	Sport & Recreation
34	A qualitative analysis of undergraduate sport management student skill and awareness development at an international sports event	John J. Miller, Zakary Mayo, Leslie Podlog	(2022)	Sport Management
35	A tourism immersion internship: Student expectations, experiences and satisfaction	Lisa Ruhanen, Richard Robinson, NoreenBreakey	(2013)	Tourism
36	An inclusive multifaceted approach for the development of electronic work-integrated learning (eWIL) curriculum	Ancy Gamage	(2022)	N/A
37	Interface: Establishing knowledge networks between higher vocational education and businesses	Joseph Kessels, Kitty Kwakman	2007	N/A

knowledge, technical (hard) and generic (soft) skills, and the context in which these are applied. If course LOs have been designed and written effectively, they will help guide the learning pathways for students and align assessments to demonstrate the intended student learning.

Bowden et al. (2000) describes graduate attributes as the qualities, skills, and understandings a university community agrees its students would desirably develop during their time at an institution. These attributes consequently shape the contribution they can make to their profession and as citizens. Bridgstock (2009) identifies two main attributes: an individual's capacity for citizenship and ability to obtain and maintain work.

Graduate Attributes represent the broad qualities necessary for students to contribute meaningfully to their communities. These attributes are underpinned by various transferable skills employers look for when recruiting and can be developed through assessment and learning experiences. Griffith University introduced graduate attributes through policy in 2016. The aim was to engender these characteristics in its graduates through bachelor's degrees and postgraduate programs. The following six GAs are available for every course for educators to embed:

1. Knowledgeable and skilled, with critical judgement

2. Effective communicators and collaborators
3. Innovative, creative, and entrepreneurial
4. Socially responsible and engaged in their communities
5. Culturally capable when working with First Australians
6. Effective in culturally diverse and international environments

Firstly, the LOs and GAs were identified for each of the 29 WIL courses offered to determine whether they share similar aims for the WIL program. With this, the LO and GA for each course were collated and established where they do or do not align. Since there are 179 LOs across the 29 WIL courses, LOs were simplified into eight general categories using text analysis. These categories stemmed from extracting the main keywords from each LO of every WIL course and grouping them by general meaning. For example, for the course, 3593QCA—Liveworm 1, LO 3 is "apply relevant competencies, technical skills, and knowledge [main keywords] applicable to professional design practice". Thus, the general learning outcome is Professional Competence (as identified in the LO categories below).

The set of LO categories included:

1. Analysis
2. Communication
3. Innovation
4. Professional Competence
5. Application of Knowledge
6. Professional Ethics
7. Collaboration
8. Multiculturalism

For the correlations, the first analysis focused on WIL type and LO, another on type and GA, and the last one studied the relationship between the WIL types, LOs, and GAs together.

Since LOs are integral to course and curriculum design, they are a valuable tool to measure and indicate WIL courses' intention and focus. As mentioned, if course LOs have been written and implemented effectively in a course, their intended outcomes will mirror the student learning outcomes for which they were designed. That is, students should demonstrate the knowledge, skills and application of these for the WIL course that they have completed. The GAs complement the learning outcomes and will support students to develop the personal attributes that employers and the broader community value. Together, LOs and GAs provide a strong coupling for designing courses and a strong indicator of the effectiveness of WIL courses across the sciences.

2.3 Educator Perspectives

The second investigation explored the viewpoints of WIL course conveners through a two-hour workshop attended by nine staff members from the Sciences Group and [name withhold] College of Arts. The workshop was divided into two parts. The first concerned a general discussion of the preliminary findings of the desktop review. In the second half, perspectives were sought regarding the challenges and potential solutions for implementing WIL. In canvassing educator perspectives, this study was able to compare and validate findings from the desktop analysis.

2.4 Equity, Diversity and Inclusion

The third investigation focused only on one discipline (architecture) and reviewed the gender, nationality, and course level of the WIL students from the inception of the architecture discipline at [name withhold] University from 2013 until 2023. The rationale for this choice was twofold. Firstly, the reviewed literature indicated that architecture had yet to be studied in this context; therefore, there was an opportunity to contribute new knowledge. Moreover, a comprehensive and complete data set was available to the research team, making a more profound analysis possible. Lastly, concerning EDI, the discipline of architecture is increasingly considering these features as fundamental principles of education and professional practices (Dos Santos, 2022; Kim, 2022; Thompson & Soccio, 2022).

2.5 Limitations of the Method

The main limitation of this study's methodology is that it focuses on a single institution. As this is the first time such a focus on WIL courses has been conducted within this institution, the findings need longitudinal data to identify and explain trends. However, the study can be extended to implement changes and monitor the impact of such changes over time and across disciplines. Another limitation of this research method is the inability to measure the intended learning outcomes compared to the actual ones. While assessment items will provide many clues, it was beyond the scope of this study to evaluate assessment items and/or interview students who had completed WIL courses. Despite these limitations, the study is strengthened by the sample size (29), which spans the breadth of various disciplines across the sciences and arts faculties.

3 Results

3.1 General WIL Context: Common Traits and Gaps

Four major findings emerged from the desktop analysis.

3.1.1 Structure

The contextual setting of WIL offering is well established at Griffith] University, as there is one structure—Partnerships Office (PO)—dedicated to supporting WIL opportunities. The PO operates as a front door for external partners and coordinates a broad range of engagement activities between our staff, students and industry partners" (https://www.griffith.edu.au/griffith-sciences/partnerships). Specifically, for WIL courses, the PO provides generic information to existing and potential industry partners to engage in WIL and information, links, and a portal for students looking to engage in WIL.

However, the information provided is quite broad and generic. Although parts apply to all WIL, it remains challenging to determine how a WIL opportunity works at a discipline level, both for an industry member willing to offer WIL or a student seeking a WIL course. Unless the WIL course concerns larger and more well-established WIL programs, such as those in the engineering department, there is a need for more relevant and detailed information regarding other disciplines. Furthermore, since PO is set up for industry engagement, more information about other types of WIL (e.g., community engagement, simulation, etc.) needs to be provided. Overall, the current WIL setting provides the big picture that could be improved with more discipline-specific refined attributes.

3.1.2 Two Predominant WIL Types and Three Common Learning Outcomes

Using the five WIL Types identified in the literature review (subsection 2.1) and course profiles to determine the type of the 29 reviewed WIL courses, the results show a strong inclination toward *Internship* and *Project* WIL types at [University withhold] (Fig. 1). Internships are the most frequent WIL type across the eight disciplines. The next most favoured type is project-based, occurring in three disciplines, and only one discipline, Law and Arts, uses service-learning WIL. Project WIL refers to working on real-live projects/strategies/briefs provided by industry/community partners (Jackson et al., 2022). No courses have been identified that provide the incubator and simulation WIL types. Also, whilst categorizing the WIL courses, it was noticed that there needed to be more clarity in defining the WIL type only with the course profile. For example, the undergraduate course "Community Internship and Partnerships for SDGs (3002LFC)" describes both an internship and service learning WIL types: 'The Community Internship course provides students with an opportunity to combine their scholarly learning with a real-world experience in a community-based organization that supports their professional and personal development.'

The analysis of the relationship between WIL types and LOs (Fig. 2) shows that no specific suite of LOs is attributed to one type. Rather, the study found that most LOs overlapped several WIL types (except for multiculturalism), with analysis, communication, and professional ethics being common LOs to the three main WIL types for the reviewed courses.

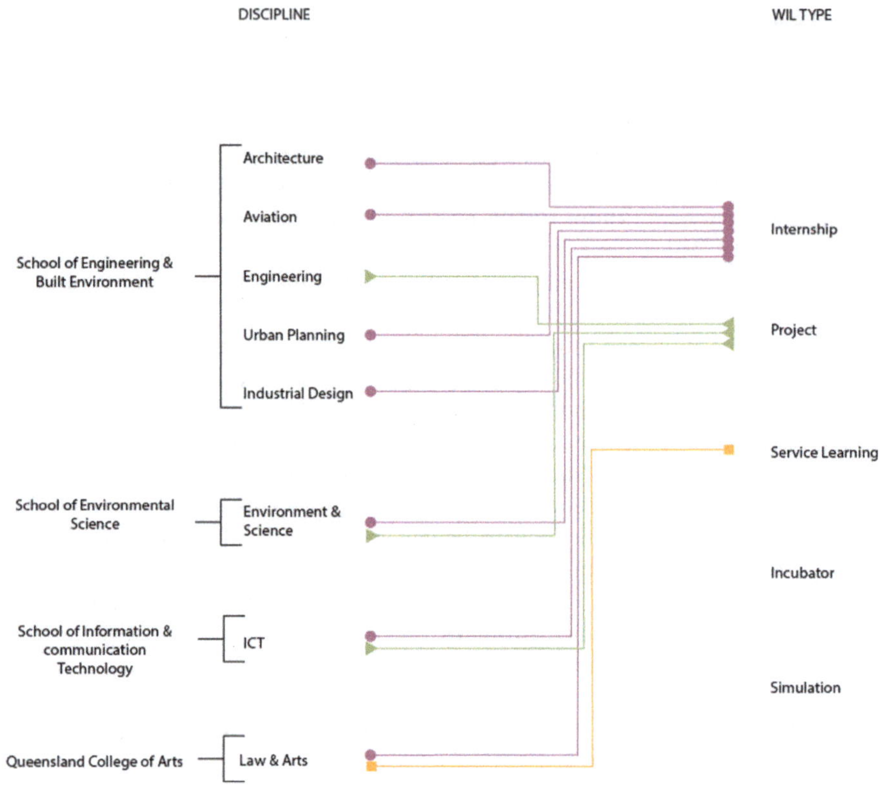

Fig. 1 WIL types by discipline

Also, the research confirmed that LOs were very disciplined and course-specific (Fig. 3).

3.1.3 Graduate Attributes: A Unique Cultural Context not Embraced

The exact process was conducted to evaluate the relationship between WIL types and Graduate Attributes. The study found that the main implemented WIL types included all six GAs at least once. However, the repetition of the GA within the 29 reviewed WIL courses shows that some GAs are preferred over others, with GAs 5 and 6 being narrowly considered in the WIL courses (Fig. 4). In particular, GA 5, "culturally capable when working with First Australians", was most underrepresented in the WIL courses.

Similarly to the LOs, the GAs across the same disciplines have a uniform GA list. For example, all ICT courses attain the same five GAs. Overall, it was found that different courses within a discipline shared a similar general LO and GA foundation. For example, the ICT discipline courses share a uniform LO and GA but differ from

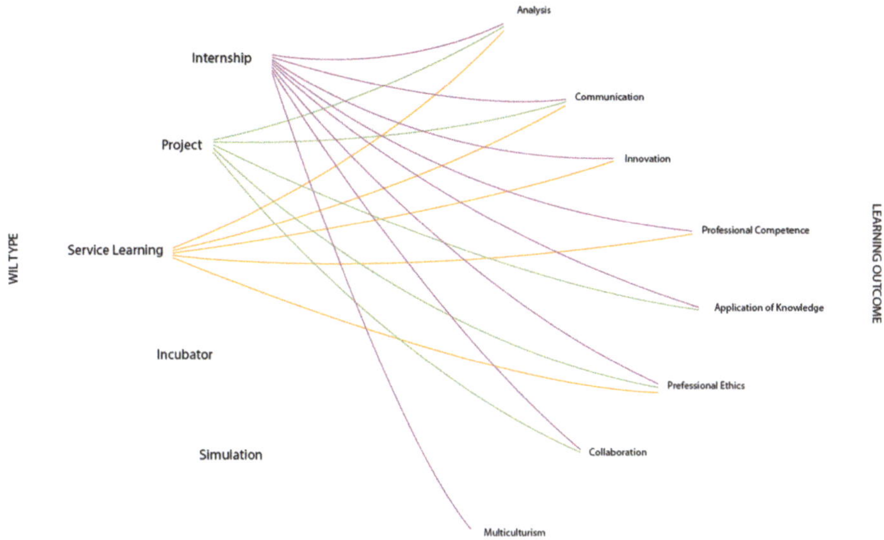

Fig. 2 The Relationship between WIL Types and Learning Outcomes

the Science discipline courses, which share their uniform LO and GA. However, there are variations in the varying specialties within the specific course. For example, course 7911ESC states broadly in the LO to "Successfully apply and further develop the knowledge and skills gained through your academic program in an external industry environment". Similarly, course 7103ESC_P1 focuses on biotechnology to "Critically analyze the application of established biotechnology project management approaches/methods to a professional industry placement or project". Despite the use of course/topic-specific LO, it was found that there is a similar relationship to the expected LO across the WIL programs.

3.1.4 Correlations WIL Types, LOs and GAs: Not Clear

Finally, it was investigated whether there is a relationship between the WIL types, LOs, and GAs. It was found that there is no clear relationship between these. It is shown (Fig. 5) that all WIL types demonstrate all six GAs despite not all LOs being utilized in each type.

By comparing the LOs against the GAs, it was found that LOs usually aligned with GAs for a particular course. However, around 70% of the WIL course LOs had at least one missing GA alignment. The most common graduate attribute discrepancy was GA 4 (socially responsible and engaged in their communities). This GA is listed in 17 of the WIL courses; however, their LOs do not corroborate this.

Other WIL relationships between LOs and GAs across disciplines were explored by comparing undergraduate and postgraduate WIL courses. However, it was found

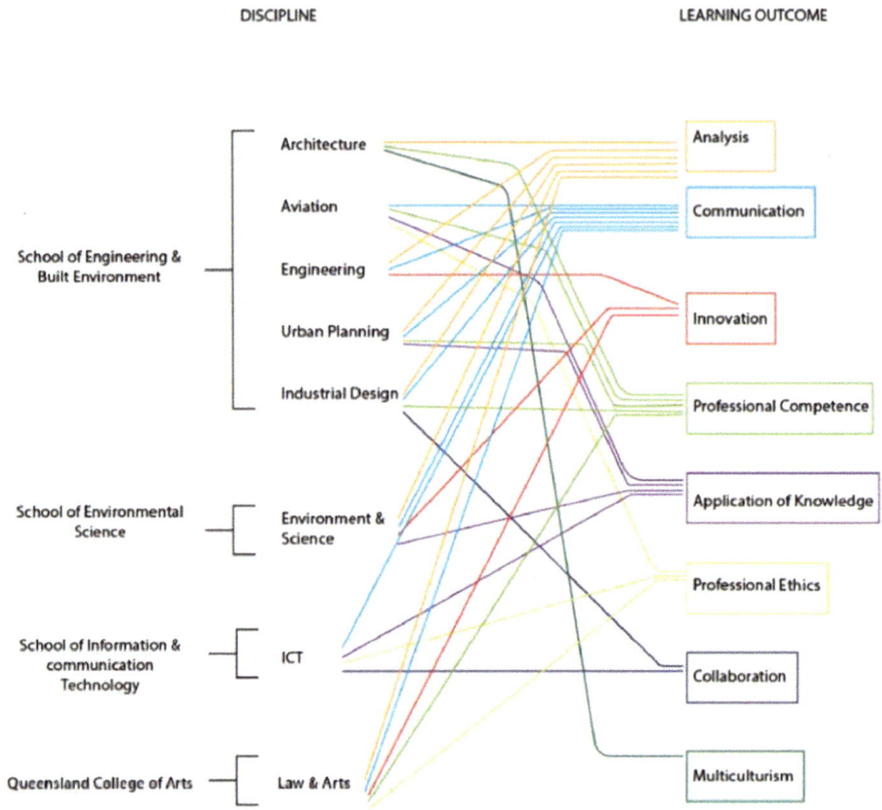

Fig. 3 The Relationship between Disciplines and Learning Outcomes

that there is no distinctive difference between a discipline's undergraduate course and a postgraduate course.

3.2 *Perspectives from the WIL Course Convenors: Challenges to Address*

After being presented with the desktop study's results, the nine WIL course convenors were invited to reflect on them and further discuss what might have been missed by the desktop study. Although nine disciplines were represented through these nine convenors, there was a clear convergence towards the challenges they expressed. Overall, three main challenges were identified.

The first one regards the administrative structure of the WIL courses that reinforced silo practices and attitudes. For example, course convenors have noted a tendency for WIL courses to function within disciplinary silos, leading to a lack

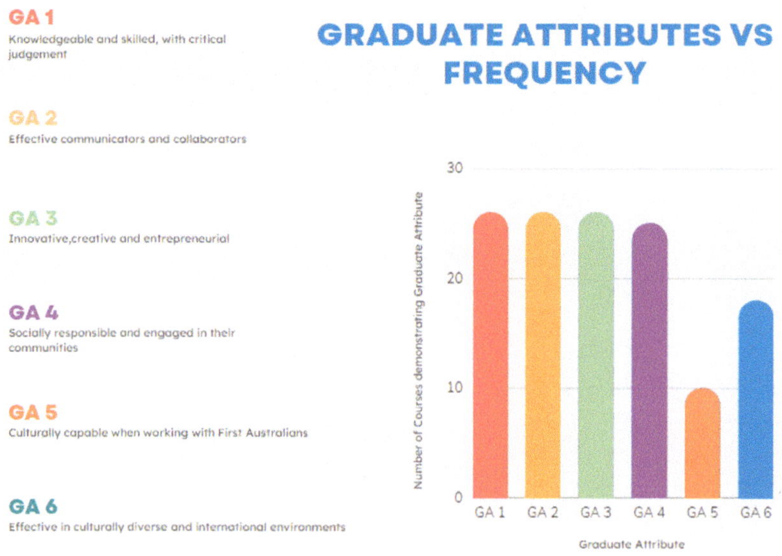

Fig. 4 Repartition of Graduate Attributes within the 29 reviewed WIL courses

of both formal and informal communication between disciplines. Furthermore, convenors often protect 'their' industry partners, with whom individual educators have spent considerable time and effort establishing a relationship. Consequently, industry partners are often aligned with specific academics rather than the department or discipline.

Another challenge concerns other aspects of the relationship with industry partners. On the one hand, ensuring the quality of industry partners, especially when the relationships are new or initiated by students, presents a challenge. While most course convenors regularly meet with students, this only sometimes extends to the industry partners. On the other hand, industry partners are also concerned with the value they can gain from offering a WIL opportunity. There is a disparity in the perceived value of students from different disciplines, with some disciplines like architecture or interior design seen as less immediately beneficial to industry partners than fields like ICT and engineering. Moreover, industry partners are often not incentivized to select international students because they offer few long-term employment prospects.

Lastly, new WIL course convenors find navigating the WIL process particularly challenging, especially in fields involving project-style WIL (such as Interior Design, Visual Art, and Architecture). This issue is exacerbated when the WIL does not fit into an established system or timeframe.

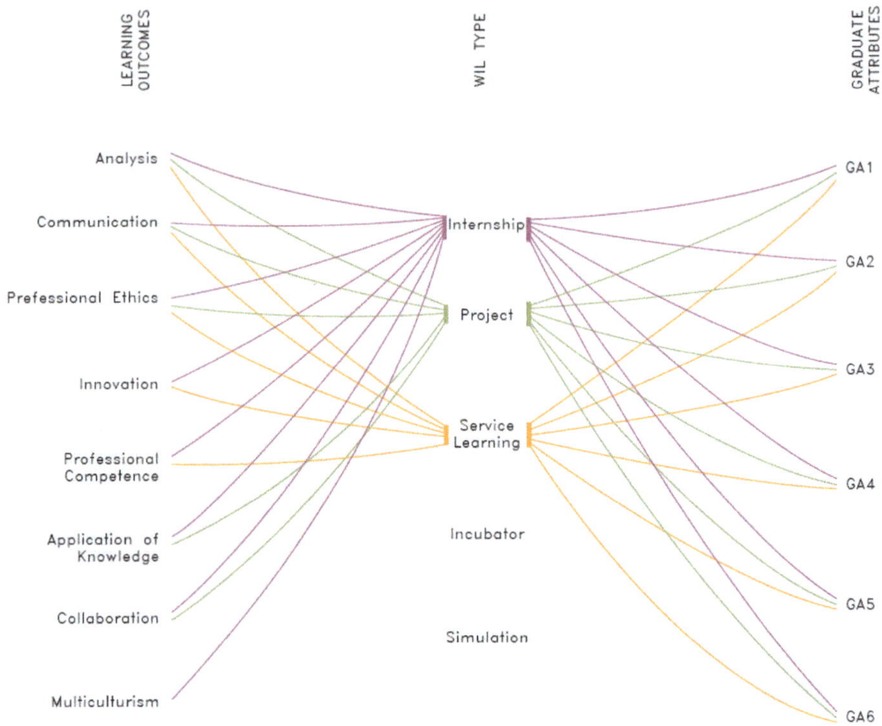

Fig. 5 LOs, WIL types and GAs

3.3 EDI: The Opportunity to Be Developed Much Further in WIL

The scrutiny of the data regarding all the 113 WIL students who undertook a WIL course in architecture from 2013 until 2023 is unequivocal: the typical WIL student is a domestic (87%) male (54%) student in Master/ Bachelor.

4 Discussion and Curriculum Design Suggestions

This study aimed to investigate whether WIL courses really work from a critical and course convenors' perspective. The following sections discuss three main results.

4.1 Curriculum Design Matters!

Although the learning and teaching scholarship has primarily investigated, analyzed and documented best practices to design a course to maximize the student experience and reflect expected values, attributes and outcomes (Bridgstock, 2009; Ferns & Lilly, 2015)), this study confirmed that the earliest step (determining LOs and GAs before designing the assessments) is of utmost importance, since incongruent curriculum design produces gaps that affect EDI engagement (e.g. LO not aligning with GA 5 regarding First Nations) and might limit the type of WIL courses being offered.

When curriculum design does matter, it can better frame expectations and delivery. Thus, one recommendation is to take the time and effort to discuss LOs and GAs to ensure they align with the expected outcomes. Frequently, course convenors' inherit' those elements and might not engage with these elements of the curriculum design (Biggs et al., 2022). This questions more generally how courses are attributed to course convenors and whether this process reflects an accurate recognition of expertise ... or convenience. Further research would certainly shed some welcome light on a phenomenon underexplored. Furthermore, since new WIL course convenors find navigating the WIL process particularly challenging, especially in fields involving project-style WIL (such as Interior Design, Visual Art, and Architecture), it could be proposed to define clear procedures at the discipline level for implementing WIL courses, including aspects like dynamic dating and modifying assessment outcomes. Moreover, where possible, formal WIL mentorships should be established.

Likewise, the administrative structure that supports WIL courses needs to be supported by enough qualified staff and means to represent the variety of WIL disciplines to enhance both students and industry partners' engagement. At a time when partnerships and authentic Learning are valued and sought after (Ferns et al., 2014), it seems crucial to have the resources to do so. This is consistent with existing scholarship evidencing how often universities struggle to achieve staff and financial threshold to meet their ambitions (Patrick et al., 2008).

Lastly, since our study showed that there was a trend that both convenors and disciplines operated in silo, leading to a lack of both formal and informal communication between disciplines and convenors, it is suggested that more informal social gatherings be organized. A specific proposal included holding a working lunch every trimester, featuring a topical presentation followed by a period for informal discussion to foster interdisciplinary ties.

To ensure the quality of industry partners, a proposed solution is to implement mandated catch-up meetings with industry partners during the project's duration, especially in disciplines where a robust system still needs to be implemented. Establishing a comprehensive database of industry partners could also contribute to a more systemic relationship with these partners.

4.2 We Need to Engage Better with Equity, Diversity and Inclusion

Our study showed the lack of EDI in architecture, consistent with the existing scholarship. Although most of the reviewed papers (see Table 1) addressed the limitation of their research due to the small sample size (Gibson et al., 2016; Marshall, 2012; Orr & Simmons, 2011; Ramage, 2014; Workman et al., 2007), those marginalized or underrepresented groups form a huge gap in current WIL programs and research. Specifically in WIL research, the literature review also found that research on WIL focuses primarily on students' degree programs and work placement locations or status of graduation or employment, particularly when examining the effectiveness of WIL in specific fields or industries. Little attention has been paid to demographic characteristics of samples or participating students, which makes it problematic to identify potential disparities or inequities in access to and outcomes of WIL experiences. However, others, such as Jackson and Collings (2018), clearly addressed excluding international students in their research samples for their first survey as they were too hard to contact.

Regarding WIL practices, common barriers to inclusivity in WIL include geographic location, financial resources, disabilities, and cultural and language barriers. For example, Maidment (2010) explains how students face financial stress caused by additional costs associated with undertaking WIL, especially when they don't have access to any form of funding. This situation is exacerbated for students in rural areas.

Another aspect of the need for more diversity is the limited variety of disciplines or bias towards specific disciplines. As the trends summarise, some academic disciplines or industries, such as health and business, offer more WIL opportunities than others, creating an uneven distribution among students. At the same time, this offers an opportunity for future research with the inclusion of other disciplines (Benati et al., 2021).

According to the National Strategy on WIL in University Education (Universities Australia et al., 2015), increasing inclusivity of WIL is a priority, and it was proposed to:

- Build support—among students, universities, employers across all sectors and governments—to increase participation in WIL
- Address equity and access issues to enable students to participate in WIL
- Increase WIL opportunities for international students and for domestic students to study off-shore.

This study showed that curriculum design is a—free—first step to address these challenges. At the same time, other proactive and collaborative strategies may include providing financial support, offering alternative WIL options, providing accommodation, addressing cultural barriers, and providing language support, which could help to ensure that all students have access to meaningful WIL experiences. Considerations for simulated or hybrid WIL offerings will improve accessibility for international students, students with disabilities, or those who need to complete regular work to support themselves, all of whom struggle to engage in off-campus WIL placements. For curriculum redesign and renewal, authentic assessment is critical. However, WIL encompasses such diversity and complexity that creating fair and equitable assessments is challenging (Ferns & Lilly, 2015).

Students from diverse backgrounds may need more social or cultural experience to integrate seamlessly into new working environments. As such, efforts to clearly define and model expectations are important (for both the industry partner and the comfort/confidence of the participating students). Preparing such students before engaging in WIL is an emerging issue addressed in the discussion. However, project management, professional behaviour and expectations, and industry/technical writing compared to academic reports are professional practices that could be considered part of pre-WIL preparation.). Based on Fujii et al. (2020), the researchers are proposing the following framework that could be implemented:

- Establish the specific needs of the student. This may include an understanding of their prior background and prior experience.
- Identify and express learning goals.
- Create a checklist that includes easy-to-understand, jargon-free learning performance statements for professional practices. Be sure to include soft skills such as client interaction, email etiquette, dress code, etc.
- Provide a one-day, in-house WIL orientation prior to starting. Here, behaviours and expectations can be expressed and modelled. This also provides an opportunity for students to confirm they understand the learning goals and ask questions.
- Regular check-in and reflections
- Formalized post-WIL check-in and reflection are wherein industry partner feedback, LO attainment, and personal reflections are discussed. This is compared to the initial learning goals. At this stage, issues are identified and would help develop better strategies to be implemented for the next cohort.

Regarding the first step of the above framework specifically, there is abundant scholarship and practice about the benefits of an Individualized Education Plan (IEP) –also known as an Individual Learning Plan- to assess the needs and background of the students and establish an accommodated plan for their learning (Cramerotti & Ianes, 2016; Harvey et al., 2020; Myara, 2022). However, most often IEP doesn't follow the students through to university, while some critics noted that alignment between IEP, goals and level of performance is sometimes lacking and the accommodation resumed to paperwork and compliance (Findley et al., 2022; Zeitlin & Curcic, 2014). In the case of WIL, we propose the adoption of such plans (after adapting and improving them), which in addition will prepare the students to become knowledgeable about their future employee rights. This would be another contribution from the universities, bridging the often-perceived gap between education and profession (Smith et al., 2024).

4.3 We Need More Research on WIL.

This research focused on a small section of the life cycle of a WIL course (the curriculum design) and evidenced several gaps that need to be further investigated. The existing scholarship also emphasized the lack of knowledge regarding the limited evaluation of the effectiveness of WIL programs, both in terms of their impact on student learning outcomes and their value to employers.

As previously identified, the critical limitation of this study is its methodology, which focuses on one institution. Since this is the first-ever study of this type with a focus on WIL courses conducted across these faculties at Griffith University, there needs to be longitudinal data to compare these findings or identify and explain trends. Further, the inability to measure the intended learning outcomes compared with the actual learning outcomes was beyond the scope of the research. Despite these limitations, the study is strengthened by its sample size (29) and breadth, encompassing a range of disciplines across the sciences and arts faculties. Notably, the study benchmarks WIL courses and provides a research methodology that can be extended to implement changes and monitor the impact of such changes over time and across disciplines.

Some studies have discussed assessment programs or proposed structures and plan to examine the outcomes of WIL programs in their chosen fields (Ferns & Lilly, 2015; Gaskell & Beaton, 2010; Scanlan, 2008) and while Vaughan and Moore (2016) specifically reviewed a workplace-based assessment tool in the health sector, there is still a lack of comprehensive and consistent evaluation across different programs and contexts. In other words, most WIL programs rely on a limited set of assessment methods, such as supervisor evaluations (Fleming & Pretti, 2019; Fynn et al., 2019), reflective essays or students' self-reflection (Benati et al., 2021; Jackson, 2017) or questionnaires (Alm et al., 2022; Falconer & Pettigrew, 2003; Jackson & Collings, 2018; Marshall, 2012; Thune & Storen, 2015). In summary, a questionnaire or survey is the most used assessment method (35% of total research), followed by interviews (19%).

Evaluation of outcomes is "of paramount importance" (Scanlan, 2008), as limited assessment or evaluation methods cannot fully capture the breadth of skills and knowledge that students develop during their WIL experiences. This can make it challenging to identify areas for improvement and ensure that WIL programs meet all stakeholders' needs. Meanwhile, consistent evaluation is necessary to identify best practices for designing and implementing effective WIL programs. Ferns and Lilly (2015) assert that "effective, efficient and equitable assessment is a key issue in implementing WIL experiences into the curriculum". Ferns and Lilly (2015) proposed an authentic assessment framework that targeted the general WIL program to overcome challenges faced by WIL evaluation. It may be time to test this framework.

5 Conclusion

Work-integrated Learning is a powerful educational approach that enhances the development of employability skills and employment opportunities for graduating students. However, WIL courses are often optional in university programs, so they are not given the due attention (from students) and appropriate resourcing (from institutions) they deserve. This research and the WIL scholarship emphasized the need to address these issues. In particular, if thoughtfully considered and designed, the WIL curriculum provides authentic learning experiences for students to develop a complex array of hard and soft skills in the workplace. As this research attests, the alignment of WIL types, Learning Outcomes, Graduate Attributes and authentic assessment is central to the WIL curriculum, which enables students to apply theoretical concepts through real-world tasks in a practical setting. Curriculum design for WIL matters, and so does equity, diversity, and inclusion. More research on WIL is required, including the effectiveness of WIL programs on student learning and employability. Improving the institutional support and communication across WIL for students, convenors and industry is encouraged. This includes accommodating student diversity, plus their individual needs with Individual Learning Plans. Preparing students prior to WIL experiences, including more EDI practices and supporting them during their placement and post-WIL, will enhance the experiences for students and industry alike. The WIL scholarship, while limited, is rich in opportunities to enhance WIL through change, including adopting an authentic assessment framework.

References

Alm, K., Beery, T. H., Eiblmeier, D., & Fahmy, T. (2022). Students' learning sustainability—implicit, explicit or non-existent: A case study approach on students' key competencies addressing the SDGs in HEI program. *International Journal of Sustainability in Higher Education, 23*(8), 60–84. https://doi.org/10.1108/ijshe-12-2020-0484

Andersson, A., Graneheim, U. H., & Nilsson, M. S. (2022). Newly-graduated nurses' work-integrated Learning: A qualitative study from an educational and occupational perspective. *Nurse Education in Practice, 59*, 103290. https://doi.org/10.1016/j.nepr.2022.103290

Bayerlein, L. (2015). Curriculum innovation in undergraduate accounting degree programmes through "virtual internships." *Education & Training, 57*(6), 673–684. https://doi.org/10.1108/ET-09-2014-0110

Baier, R., & Tao, Y. (2007). *An interdisciplinary pedagogical teaching approach for engineering, in conjunction with architecture and construction with solar decathlon project.* https://doi.org/10.18260/1-2--2921

Benati, K., Lindsay, S., & Fischer, J. (2021). Applying theory in practice: Views of graduating business students. *Education + Training, 63*(9), 1213–1224. https://doi.org/10.1108/et-07-2020-0197

Berndtsson, I., Dahlborg, E., & Pennbrant, S. (2020). Work-integrated learning as a pedagogical tool to integrate theory and practice in nursing education—An integrative literature review. *Nurse Education in Practice, 42*. https://doi.org/10.1016/j.nepr.2019.102685

Biggs, J., Tang, C., & Kennedy, G. (2022). Teaching for quality learning at university 5e. Mc-Graw-Hill.

Bowden, J., Hart, G., King, B., Trigwell, K., & Watts, O. (2000). *Generic capabilities of ATN university graduates.*

Brown, B. A., Roodhouse, S., Harte, J., & Warnes, A. M. (2007). Developing the health care workforce: A comparison of two work-based learning models. *Education + Training, 49*(3), 193–200.

Bridgstock, R. (2009). The graduate attributes we have overlooked: Enhancing graduate employability through career management skills. *Higher Education Research and Development, 28*(1), 31–44.

Cramerotti, S., & Ianes, D. (2016). An Ontology-based system for building individualized education plans for students with special educational needs. *Procedia—Social and Behavioral Sciences, 217*, 192–200. https://doi.org/10.1016/j.sbspro.2016.02.062

Dean, S., Ahchay, D., Walker, F., & Ryan, C. (2022). Innovation in Australian nursing student work integrated Learning during a pandemic. *Nurse Education in Practice, 61*, 103334. https://doi.org/10.1016/j.nepr.2022.103334

Edziwa, X., & Blignaut, S. (2022). Graduate employability skills: The voice of Agricultural Technical Vocational Education and Training (ATVET) students in Zimbabwe. *South African Journal of Higher Education, 36*(2). https://doi.org/10.20853/36-2-4501

English, P., de Villier Scheepers, M. J., Fleischman, D., Burgess, J., & Crimmins, G. (2021). Developing professional networks: The missing link to graduate employability. *Education & Training, 63*(4), 647–661. https://doi.org/10.1108/ET-10-2020-0309

Falconer, S., & Pettigrew, M. (2003). Developing added value skills within an academic programme through work-based Learning. *International Journal of Manpower, 24*(1), 48–59. https://doi.org/10.1108/01437720310464963

Ferns, S., & Lilly, L. (2015). Driving institutional engagement in WIL: Enhancing graduate employability. *Journal of Teaching and Learning for Graduate Employability, 6*(1), 126–143.

Ferns, S., Russell, C., Smith, C., & Cretchley, P. (2014). *The impact of work integrated learning on student work-readiness.* Office of Learning and Teaching.

Findley, J. A., Ruble, L. A., & McGrew, J. H. (2022). Individualized education program quality for transition age students with autism. *Research in Autism Spectrum Disorders, 91.*

Fleming, J., & Pretti, T. J. (2019). The impact of work-integrated learning students on workplace dynamics. *Journal of Hospitality, Leisure, Sport & Tourism Education, 25.* https://doi.org/10.1016/j.jhlste.2019.100209

Fujii, R., McElhaney, K., & Alozie, N. (2020). Promoting equity and inclusion in STEM curriculum design. *ICLS Proceedings.* https://doi.org/10.22318/ICLS2020.1747

Fynn, A. W. R., Dladla, S. N., & Erasmus, H. (2019). A study on the skills of master of arts in research psychology graduates from the employers' perspective. *South African Journal of Higher Education, 33*(4). https://doi.org/10.20853/33-4-3397

Gamage, A. (2022). An inclusive multifaceted approach for the development of electronic work-integrated learning (ewil) curriculum. *Studies in Higher Education, 47*(7), 1357–1371. https://doi.org/10.1080/03075079.2021.1894116

Gaskell, L., & Beaton, S. (2010). Inter-professional work-based Learning within an MSc in Advanced Practice: Lessons from one UK higher education programme. *Nurse Education in Practice, 10*(5), 274–278. https://doi.org/10.1016/j.nepr.2009.11.017

Gibson, T., Kerr, D., & Fisher, R. (2016). Accelerating supply chain management learning: Identifying enablers from a university-industry collaboration. *Supply Chain Management: An International Journal, 21*(4), 470–484. https://doi.org/10.1108/scm-10-2014-0343

Glavas, C., & Schuster, L. (2020). Design principles for electronic work integrated Learning (eWIL). *The Internet and Higher Education, 47.* https://doi.org/10.1016/j.iheduc.2020.100760

Harvey, J., Farquharson, K., Schneider-Cline, W., Bush, E., & Pelatti, C. Y. (2020). Describing the composition of individualized education plans for students with Traumatic brain injury. *Language, Speech, and Hearing Services in Schools, 51*(3), 839–851.

Jackson, D. (2017). Developing pre-professional identity in undergraduates through work-integrated Learning. *Higher Education, 74*(5), 833–853. https://doi.org/10.1007/s10734-016-0080-2

Jackson, D., & Collings, D. (2018). The influence of Work-Integrated Learning and paid work during studies on graduate employment and underemployment. *Higher Education, 76*(3), 403–425. https://doi.org/10.1007/s10734-017-0216-z

Jackson, D., & Wilton, N. (2016). Developing career management competencies among undergraduates and the Role of work-integrated Learning. *Teaching in Higher Education, 21*(3), 266–286.

Jackson, D., Riebe, L., & Macau, F. (2022). Determining factors in graduate recruitment and preparing students for success. *Education + Training, 64*(5), 681–699. https://doi.org/10.1108/ET-11-2020-0348

Kessels, J., & Kwakman, K. (2007). Interface: Establishing knowledge networks between higher vocational education and businesses. *Higher Education: THe International Journal of Higher Education and Educational Planning, 54*(5), 689–703. https://doi.org/10.1007/s10734-006-9018-4

Kim, J. (2022). Making architecture relevant to underserved communities: mapping reconsidered. *Architecture.* https://doi.org/10.3390/architecture2040034

Maidment, J. (2010). Problems experienced by students on field placement: Using research findings to inform curriculum design and content. *Australian Social Work, 56*(1), 50–60. https://doi.org/10.1046/j.0312-407X.2003.00049.x

Marshall, J. E. (2012). Developing midwifery practice through work-based Learning: An exploratory study. *Nurse Education in Practice, 12*(5), 273–278. https://doi.org/10.1016/j.nepr.2012.06.003

Miller, J. J., Mayo, Z., & Podlog, L. (2022). A qualitative analysis of undergraduate sport management student skill and awareness development at an international sports event. *Journal of Hospitality, Leisure, Sport & Tourism Education, 30.* https://doi.org/10.1016/j.jhlste.2021.100345

Moroney, T., Gerdtz, M., Brockenshire, N., Maude, P., Weller-Newton, J., Hatcher, D., Molloy, L., Williamson, M., Woodward-Kron, R., & Molloy, E. (2022). Exploring the contribution of clinical placement to student learning: A sequential mixed methods study. *Nurse Education Today, 113*, 105379. https://doi.org/10.1016/j.nedt.2022.105379

Myara. N. (2022). On an international level, what are the functions of an Individualized Education Program (IEP) or Transition Plan (TP)? *Revista de Psicología, 40*(1). https://doi.org/10.18800/psico.202201.006

Nagarajan, S., & Edwards, J. (2014). Is the graduate attribute approach sufficient to develop work-ready graduates? *Journal of Teaching and Learning for Graduate Employability, 5*(1).

Dentressangle, N. (2010). Norbert Dentressangle delivers managers of tomorrow. *Human Resource Management International Digest, 18*(1), 20–22. https://doi.org/10.1108/09670731011016152

Orr, K., & Simmons, R. (2011). Restrictive practice: The work-based learning experience of trainee teachers in English further education colleges. *Journal of Workplace Learning, 23*(4), 243–257.

Patrick, C. J., Peach, D., Pocknee, C., Webb, F., Fletcher, M., & Pretto, G. (2008). *The WIL (Work Integrated Learning) report: A national scoping study.* Australian Learning and Teaching Council.

Peach, D., Gomez, R., & Ruinard, E. (2013). Reconstructing spaces and places through blended work integrated Learning. *Research and Development in Higher Education: THe Place of Learning and Teaching, 36*, 336–345.

Prinsley, R., & Baranyai, K. (2015). *STEM-trained and job-ready* (Occasional Paper series, 12). Office of the Chief Scientist: Australian Government.

Ramage, C. (2014). Learning to learn through university accredited work-based Learning: A threshold conception. *Journal of Workplace Learning, 26*(8), 488–510. https://doi.org/10.1108/jwl-06-2013-0042

Ruhanen, L., Robinson, R., & Breakey, N. (2013). A tourism immersion internship: Student expectations, experiences and satisfaction. *Journal of Hospitality, Leisure, Sport & Tourism Education, 13*, 60–69. https://doi.org/10.1016/j.jhlste.2013.02.001

Santos, J. (2022). Using appreciative inquiry to develop organizational values rooted in social justice and diversity, equity and inclusion. *AI Practitioner*. https://doi.org/10.12781/978-1-907549-53-3-3.

Savic, M., & Kashef, M. (2013). Learning outcomes in affective domain within contemporary architectural curricula. *International Journal of Technology and Design Education, 23*, 987–1004. https://doi.org/10.1007/s10798-013-9238-8

Scanlan, J. N. (2008). A theoretical model for supporting bounded learning communities focussed on the promotion of clinical reasoning in health students undertaking field-based education. *Focus on Health Professional Education, 10*(1), 27–42.

Smith, S., Dupre, K., & Crough, J. (2024). Beyond the hard and soft skills paradigm: An Australian architecture industry perspective on employability and the university/practice divide. *Higher Education, Skills and Work-Based Learning.* https://doi.org/10.1108/HESWBL-07-2023-0197

Snyman, L. W., & Kruger, D. (2020). New models for improving teaching and training in engineering and technology transfer in South Africa. *South African Journal of Higher Education, 34*(6). https://doi.org/10.20853/34-6-3998

Swart, C. (2014). An assessment of work-integrated Learning for public relations in an open distance learning context. *Public Relations Review, 40*(2), 387–396. https://doi.org/10.1016/j.pubrev.2013.10.008

Talbot, J., & Roodhouse, S. (2007). Delivering distance education for modern government: the F4Gov programme. *Education + Training, 49*(3), 250–260. https://doi.org/10.1108/00400910710749387

Thompson, J., & Soccio, P. (2022). 'This is what gets people hired!': Academic perspectives on employability skills in architecture and the potential impact of COVID-19. *Journal of Teaching and Learning for Graduate Employability.* https://doi.org/10.21153/jtlge2022vol13no1art1555

Thune, T., & Støren, L. A. (2015). Study and labour market effects of graduate students' interaction with work organizations during education. *Education + Training, 57*(7), 702–722. https://doi.org/10.1108/et-10-2014-0126

Universities Australia. (2015). *National Strategy on WIL in University Education.* Universities Australia.

Vaughan, B., & Moore, K. (2016). The mini Clinical Evaluation Exercise (mini-CEX) in a pre-registration osteopathy program: Exploring aspects of its validity. *International Journal of Osteopathic Medicine, 19*, 61–72. https://doi.org/10.1016/j.ijosm.2015.07.002

Walters, D., & Zarifa, D. (2008). Earnings and employment outcomes for male and female post-secondary graduates of coop and non-coop programmes. *Journal of Vocational Education & Training, 60*(4), 377–399. https://doi.org/10.1080/13636820802591863

Workman, B., & Costley, C. (2007). Casing the joint. *Journal of Workplace Learning, 19*(3), 146–160. https://doi.org/10.1108/13665620710735620

Zeitlin, V. M., & Curcic, S. (2014). Parental voices on individualized education programs: 'Oh, IEP meeting tomorrow? Rum tonight!' *Disability & Society, 29*(3), 373–387. https://doi.org/10.1080/09687599.2013.776493

Chapter 9
Industry Evaluation of Students' Skills and Competencies

Ivan Gratchev, Hugo G. Espinosa, and Andrew Busch

Abstract The School of Engineering and Built Environment at Griffith University in Queensland, Australia, offers a unique Industry Affiliates Program (IAP) as part of its Work Integrated Learning (WIL). The program provides students from various engineering disciplines with opportunities to work on real-life engineering projects in the industry for one trimester. Students must possess specific knowledge and skills before entering industry placements; however, teaching every necessary skill at a university may be challenging. As a result, mismatches between the student knowledge and skills and the industry needs may still exist. Regular communications between the university and industry sector must be established to address this issue, providing feedback on the type of skills and competencies required by the industry. This chapter presents and discusses such a university-industry relationship that evaluates student performance during their industry placements. An online survey of 207 industry partners and qualitative data analysis found that most students who undertook industry placements demonstrated a high level of knowledge, skills, and competencies that led to successful project completion. However, despite the relatively high level of industry satisfaction, a few mismatches between students' and industry partners' perceptions of the industry placement were identified and discussed to identify areas for improvement. The results and lessons learned would improve the WIL curriculums and assist with recommendations on better supporting students in acquiring the knowledge and developing critical skills required by the relevant industry.

I. Gratchev (✉)
Griffith University, Gold Coast, QLD, Australia
e-mail: i.gratchev@griffith.edu.au

H. G. Espinosa · A. Busch
Griffith University, Nathan, QLD, Australia

I. Gratchev et al. (eds.), *Transformative Approaches to Career-Ready Education*,
https://doi.org/10.1007/978-981-96-7018-5_9

1 Introduction

Universities are essential in creating national wealth by providing skilled labour to support productive efforts (Smith, 2016a). To produce work-ready graduates, universities have adopted strategies that incorporate work-integrated learning (WIL), a broad term for various activities designed to provide students with exposure to real-world experience by integrating theory with work practice (Patrick et al., 2009). Although WIL can be incorporated into classroom learning (Kramer & Usher, 2012), it mainly refers to educational programs that include a work placement. Such industry placements allow students to develop skills and competencies expected by the industry and prepare them for entering the workforce. This introduction will briefly discuss the changes in engineering education in Australia over the past decades, focusing on developing industry-related generic skills and competencies.

In the early 1990s, Australian universities were required to produce graduates with the skills, personal attributes, and values regardless of students' discipline or field of studies (Young & Chapman, 2010). The Government emphasized the importance of the quality of university education by creating the Australian Universities Quality Agencies (https://www.teqsa.gov.au), which conducted quality audits of universities. As a result, the Course Experience Questionnaire CEQ (Ramsden, 1991) was introduced and became part of Australia's quality and funding regime. The CEQ was later extended by adding a scale to measure generic skills outcomes (Wilson et al., 1997). In addition to technical skills, Engineers Australia, an Australian national body for engineering, introduced a notion that all curricula should expose the production of 'soft skills' (or 'professional skills') outcomes for engineering graduates.

Later, due to economic growth, the Australian Government began to emphasize the importance of generic attributes that students in higher education should develop so that they can solve social and economic problems (West, 1998). The economy produced a demand for employees with excellent generic competencies (not only subject-specific technical skills). Richens and McClain (2000) reported the results of a study in which 400 Australian employers from different industry sectors indicated the importance of generic competencies such as sound interpersonal awareness for workplace success. Owen et al. (2000) reported that 'soft/professional skills' such as communication and interpersonal style were among the most critical competencies expected by the industry. Crowley et al. (2000) interviewed industry partners from the construction industry and reported that communication, teamwork, and planning were essential competencies required from employees.

Further industry demands and discussions about the values of educational outcomes focused on developing generic skills, and the language changed towards 'graduate attributes' (Barrie, 2007), leading to further changes in engineering education in Australia. These changes were defined mainly by outcomes required by Engineers Australia for program accreditation (Engineers Australia (EA) 2005 a, b, QAA 2006, Accreditation Board for Engineering Technology (ABET) 2008), which stipulated ten generic attributes, with half of the attributes being non-technical.

During this time, several medium and large-scale studies (Bodmer et al., 2002; Brumm et al., 2006; Ferguson, 2006; Lang et al., 1999; Male et al., 2011a, 2011b; Male 2010; Meier et al., 2000; Nair et al., 2009; Nguyen, 1998; Scott & Yates, 2002; Spinks et al., 2006) were conducted to identify the critical skills and competencies expected from graduate students in Australia and other countries, including the USA and Europe. These studies that involved interviewing industry partners and employers produced similar results, emphasizing that non-technical attributes, such as communication, teamwork, and attitude, were as important in engineering practice as technical skills. The attitudinal components were honesty, self-motivation, demeanour, creativity, and concern for others.

In the past decades, the focus has been mainly on the link between graduate attributes and their relevance for employability, causing higher education institutions to engage more deeply with the WIL agenda (Smith et al., 2016b; Smith, 2012). Although the critical attributes and competencies were identified, there still needed to be more debate over the university's role in developing such skills, with a major concern related to the lack of relevant industry-related project work and emphasis on rote learning and memorization. Recently, Ajjawi et al. (2020) reported that there might be mismatches between what students were taught at the university and how it was practised at the workplace.

It is essential to ensure that the knowledge and skills students develop at a university can be readily applied in an industry practice and meet the industry's needs (Kensington-Miller et al., 2018). To achieve this, educators should receive regular industry feedback that helps refine curricular and educational programs to accommodate industry demands and new trends. The experience indicates that this task can be technically challenging, requiring significant time and effort from the university and the industry sector. This chapter presents and discusses an example of collaboration between Griffith University and industry, which offers engineering students industry placements in the final year of their studies.

2 Research Questions

Griffith School of Engineering and Built Environment offers an Industry Affiliates Program (IAP) to provide undergraduate and master students across various engineering disciplines (i.e., civil, environmental, mechanical, electrical, electronic, software, and project/construction management) with opportunities to work on real-life engineering projects in the industry for one teaching period (12-week trimester). Students undertake an industry placement and complete an industry-based thesis to gain work experience before graduation. This program is offered across three trimesters, and all industry placements are arranged and managed by the Partnership Office. Students can secure an industry placement through the Griffith Science Partnership Office, which offers several industry projects every trimester, or by sourcing industry placements by themselves, often linked to their previous internships. This program has been run over several years. As a result, the Partnership Office has established contacts with industry partners, mainly from Brisbane and the Gold Coast area of Queensland, Australia who are willing to offer industry placement/s to students every year.

Given the current knowledge, the research questions of this study are as follows:

1. Are industry partners satisfied with the knowledge and skills of graduating university students?
2. What generic skills and competencies do university students possess (or lack) before graduation?

Through a series of online surveys, this study seeks to identify how industry partners evaluate the level of skills and competencies of Griffith University's final-year students and to identify areas for improvement in current engineering programs.

3 Methodology

The study consisted of an online questionnaire of industry partners conducted at the end of student industry placements. Although no incentives were offered for participation, many industry partners voluntarily completed the survey because by doing so, they could provide feedback about student performance. An invitation to complete an online questionnaire was emailed to all industry partners in Week 11, and the survey was open for two weeks. This was to collect industry partners' evaluation of student performance and their comments about student engagement, project management, and achievements. The following questions were asked:

Q1. How would you rate the ethical and professional conduct of the student throughout the placement?
Q2. How would you rate the student's overall management of the project/placement?
Q3. How would you rate the delivery of project/placement outcomes?
Q4. How would you rate the student's use of initiative and responsibility in the workplace?
Q5. How would you rate the student's reliability, dependability, and integrity?

Each question had a Likert scale of answers with attached points, ranging from the lowest of 'Unacceptable' with 2 points through 'Some difficulties exist' with 4 points, 'Satisfactory performance' with 6 points, 'Good performance' with 8 points, to the highest of 'Exceptional performance' with 10 points. The points for each question were summed up, and the total mark (out of 50) was used to evaluate student performance quantitatively. The higher the score, the greater the satisfaction of industry partners with student performance. The industry partners also commented on each question and provided a rationale for their decision.

All the data was summarised in an Excel® spreadsheet. The points for all five answers were added up and used to measure the level of industry partner satisfaction with student performance.

The industry comments were analyzed using Leximancer (www.leximancer.com/), a commercially available text and data analysis software. The approach employed a thematic analysis and identified major concepts in each theme. The concepts assisted with understanding industry satisfaction with student performance, identifying students' skills and competencies, and potential areas for improvement. For each identified theme, industry partners' quotations that provided insights into student performance were extracted and analyzed.

4 Results and Discussion

Two hundred-seven responses were collected from industry partners over two trimesters. The total marks are summarized in Fig. 1 as the distribution of total marks against the number of students who received this mark. This data indicates that many industry partners were delighted with students' performance, giving them high marks. For example, more than 70 students received the highest mark from their industry partners. The experience showed that students who received such high marks established effective and productive relationships with their industry partners and completed their projects on time with high engagement and initiative. This high evaluation also suggested that the industry partners were satisfied with the knowledge and skills that students had developed at university. However, it is noted that during the placement, the students could establish close relationships with their industry partners, which might have led to bias that the industry partners possessed during their evaluation of student performance.

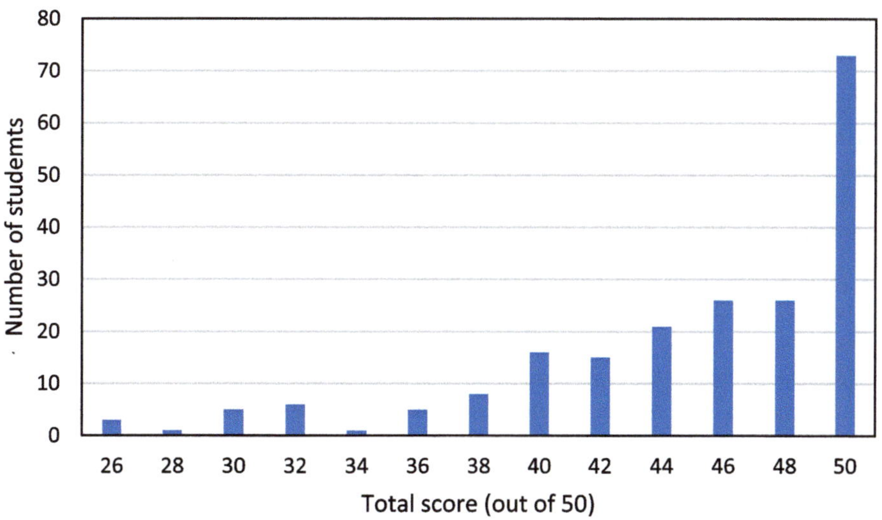

Fig. 1 Distribution of total marks among the students who undertook industry placements

Only a few students received relatively low total marks (below 40), mostly because they failed to properly manage certain aspects of their projects. This could be due to a lack of certain skills, competencies, attitudes, or professional conduct.

Analysis of the qualitative feedback from the industry partner survey involved the identification of themes and related concepts, as graphically shown in Fig. 2. The major themes were project (462 hits), work (387 hits), and time (385 hits). Each theme included concepts, as shown in Table 1, highlighting different student performance attributes. For example, the 'project' theme included initiative, supervision, student, and level. The analysis indicated that students often showed initiative at a high level during their projects. A few industry partners' quotations given below elaborate on this point:

'[Student Name] consistently showed initiative through the sharing of ideas'.

'This level of initiative is definitely something that employers are looking for'.

'[Student Name] displayed a high level of initiative and assumed an appropriate amount of responsibility for the outcome of this project'.

A few comments indicated that students required a little supervision from their industry partners as the students knew well what they were expected to achieve.

[Student Name] required minimal direction throughout the project, effectively taking the project lead and bringing to a successful conclusion.

The 'work' theme consisted of the following concepts: 'able, showed, and demonstrated', which suggested that the industry partners were satisfied with the work that students demonstrated during their placements. The quotations included:

She had a great work ethic and always showed great respect for work and the people she worked with.

The student showed more advanced skills for project undertaking and investigation and demonstrated a higher-degree approach to project organization and reporting.

The other main themes were teamwork, professionalism, and skills. Many industry partners commended their students for teamwork skills, professionalism in the workplace, and positive attitudes and ethics. In general, industry partners indicated that the students exceeded their expectations. The following quotations provide typical examples of industry partner feedback:

'[Student Name] has been a good fit to the team.'

[Student Name] has been very professional through the learning process and industry engagement.

[Student Name] was highly professional, always punctual and attentive.

All exchanges have been extremely ethical and held to high professional standards by [Student Name].

Industry partners expressed an opinion that students could learn on-site when provided with specific training at a workplace:

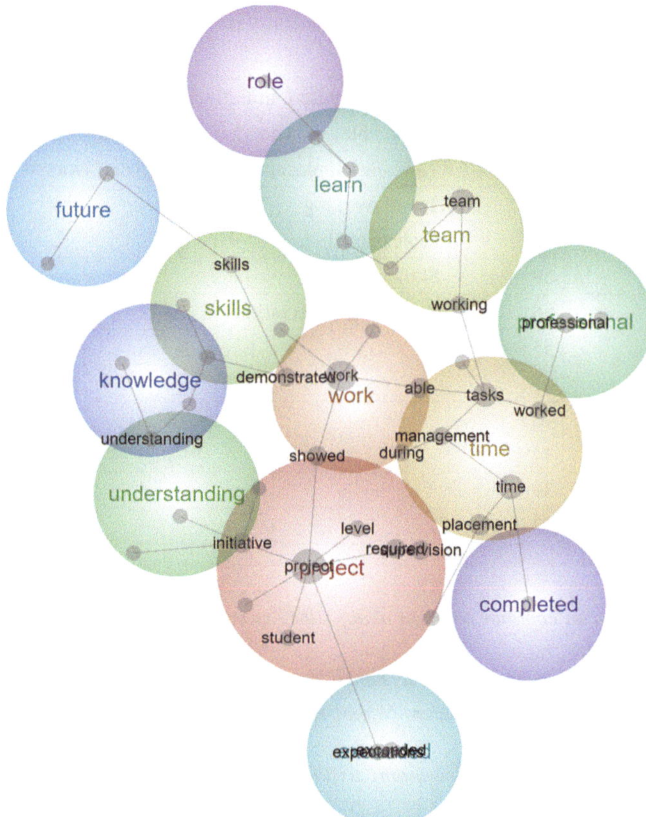

Fig. 2 Major themes and concepts derived from industry partners' feedback

> *[Student Name] was always keen to listen, learn and put into practice teachings for positive results with our procedures and processes.*
>
> *Very fast learner and incredibly motivated in everything he does.*

Such industry placements benefit students who gain industry experience and apply their knowledge to practice and the industry, which can select bright students, train them, and offer them a job after graduation.

> *We will be happy to have [Student Name] return post-university.*
>
> *'she became a trusted member of our team'.*

Challenges. Although a large cohort of students demonstrated sufficient knowledge and skills required by the industry, there were still attributes and competencies for students to learn and develop. One of the advantages of such industry placements was that they showed what skills students should develop to become successful graduate engineers.

Table 1 Summary of themes and concepts from industry partners' feedback

Theme	Hits	Concepts
Project	462	Project, required, initiative, supervision, student, level
Work	387	Work, able, showed, demonstrated, during
Time	385	Time, tasks, placement, management, worked
Team	217	Team, working
Professional	109	Professional
Skills	59	Skills
Exceeded	56	Exceeded, expectations
Future	55	
Understanding	51	
Learn	46	
Completed	44	
Knowledge	39	
Role	27	

Hopefully, you can continue to learn about BIM, as the knowledge you currently have is very basic.

It would be good to see more of the technical knowledge acquired at university.

I would like to recommend [Student Name] to continue developing his strategic thinking skills.

As discussed, please focus on developing writing skills and perhaps look into short courses for writing reports and minutes.

5 Concluding Remarks

An online survey was conducted among industry partners to obtain their evaluation of final-year Griffith University student performance during industry placements. Two hundred seven responses were collected and analyzed to identify the skills and competencies that the students possessed and potential mismatches between them and industry expectations. The major conclusions are as follows:

- Many industry partners were satisfied with student performance during their industry placements and commended students on the development of skills and competencies expected by the industry from graduate students. Many students exhibited high levels of engagement with the project, project management skills, responsibility, and initiative.
- This high level of satisfaction expressed by the industry partners indicated that the current educational programs at the Griffith School of EBE assisted students in developing generic skills and attributes expected by the industry. The final-year students could solve real-life engineering problems or learn quickly on-site to

acquire subject-specific technical knowledge and skills required by an industry sector.

– Despite the overall satisfaction, industry partners' feedback suggested that there was still room for improvement for students, which included the enhancement of technical knowledge, writing and critical thinking skills.

Acknowledgements This study was conducted in accordance with ethical standards managed by the Griffith University Ethics Committee (Ref No: 2023/109).

References

Ajjawi, R., Tai, J., Huu Nghia, T. L., Boud, D., Johnson, L., & Patrick, C. J. (2020). Aligning assessment with the needs of work-integrated learning: The challenges of authentic assessment in a complex context. *Assessment & Evaluation in Higher Education, 45*(2), 304–316.

Barrie, S. C. (2007). A conceptual framework for the teaching and learning of generic graduate attributes. *Studies in Higher Education, 32*(4), 439–458.

Bodmer, C., Leu, A., Mira, L., & Rutter, H. (2002). *SPINE: Successful Practices in International Engineering Education: Engineers Shape our Future IngCH*.

Brumm, T. J., Hanneman, L. F., & Mickelson, S. K. (2006). Assessing and developing program outcomes through workplace competencies. *International Journal of Engineering Education, 22*(1), 123–129.

Crowley, S., Garrick, J., & Hager, P. (2000). Constructing work: The hidden impact of generic competencies in the Australian construction industry. In Future Research: Research futures conference, Australian Vocational Education and Training Research Association, Canberra.

Ferguson, C. (2006). Defining the Australian Mechanical Engineer. *European Journal of Engineering Education, 31*(4), 471–485.

Kensington-Miller, B., Knewstubb, B., Longley, A., & Gilbert, A. (2018). From invisible to SEEN: A conceptual framework for identifying, developing and evidencing unassessed graduate attributes. *Higher Education Research & Development, 37*(7), 1439–1453.

Kramer, M., & Usher, A. (2012). *Work-integrated learning and career-ready students: Examining the evidence*. Higher Education Strategy Associates.

Lang, J. D., Cruse, S., McVey, F. D., & McMasters, J. (1999). Industry expectations of new engineers: A survey to assist curriculum designers. *Journal of Engineering Education, 88*(1), 43–51.

Male, S. A. (2010). Generic engineering competencies: A review and modelling approach. *Education Research and Perspectives, 37*(1), 25–51.

Male, S. A., Bush, M. B., & Chapman, E. S. (2011a). An Australian study of generic competencies required by engineers. *European Journal of Engineering Education, 36*(2), 151–163.

Male, S. A., Bush, M. B., & Chapman, E. S. (2011b). Understanding generic engineering competencies. *Australasian Journal of Engineering Education, 17*(3), 147–156.

Meier, R. L., Williams, M. R., & Humphreys, M. A. (2000). Refocusing our efforts: Assessing non-technical competency gaps. *Journal of Engineering Education, 89*(3), 377–385.

Nair, C. S., Patil, A., & Mertova, P. (2009). Re-engineering graduate skills: A case study. *European Journal of Engineering Education, 34*(2), 131–139.

Nguyen, D. Q. (1998). The essential skills and attributes of an engineer: A comparative study of academics, industry personnel and engineering students. *Global J. of Engng. Educ, 2*(1), 65–75.

Owen, G., Shelton, E., Stevens, A. B., Nelson-Christinedaughter, J., Roy, C., & Heineman, J. (2000). Whose job is it? Employers' views on welfare reform (No. 184). Northwestern University/University of Chicago Joint Center for Poverty Research.

Patrick, C. J., Peach, D., Pocknee, C., Webb, F., Fletcher, M., & Pretto, G. (2009). *The WIL report: Work integrated learning*. Queensland University of Technology.
Ramsden, P. (1991). A performance indicator of teaching quality in higher education: The course experience questionnaire. *Studies in Higher Education, 16*(2), 129–150.
Richens, G. P., & McClain, C. R. (2000). Workplace basic skills for the new millennium. *Journal of Adult Education, 28*(1), 29.
Scott, G., & Yates, K. W. (2002). Using successful graduates to improve the quality of undergraduate engineering programmes. *European Journal of Engineering Education, 27*(4), 363–378.
Smith, C. (2012). Evaluating the quality of work-integrated learning curricula: A comprehensive framework. *Higher Education Research & Development, 31*(2), 247–262.
Smith, C. D. (2016a). The emergence and development of Work-Integrated Learning (WIL): Implications for assessment, quality and quality assurance in higher education. *Reforming Learning and Teaching in Asia-Pacific Universities: Influences of Globalized Processes in Japan, Hong Kong and Australia*, pp. 337–364.
Smith, C., Ferns, S., & Russell, L. (2016b). Designing work-integrated learning placements that improve student employability: Six facets of the curriculum that matter. *Asia-Pacific Journal of Cooperative Education, 17*(2), 197–211.
Spinks, N., Silburn, N., & Birchall, D. (2006). *Educating engineers for the 21st century: The industry view*. The Royal Academy of Engineering.
West, R. (1998). *Learning for life: Review of higher education financing and policy*. Australian Government Publishing Service.
Wilson, K. L., Lizzio, A., & Ramsden, P. (1997). The development, validation and application of the course experience questionnaire. *Studies in Higher Education, 22*(1), 33–53.
Young, J., & Chapman, E. (2010). Generic competency frameworks: A brief historical overview. *Education Research and Perspectives, 37*(1), 1–24.

Chapter 10
Future-Ready Leadership: Insights from Industry

Ramabadran Gopalakrishnan

Abstract The style of this chapter may appear different from other chapters insofar as I write in the first person. This is because i am sharing professional insights from a non-academic domain, and the insights are personal. My exposure to academic leadership has only been tangential through my membership in the governing boards of a few management institutes over the last 25 years. All of my career has been in the corporate sector. However, some 'Brahma mantras' of leadership apply to all fields, irrespective of the domain. In short, there are universal lessons for future leaders, and the accompanying insights should interest readers. In summary, given rapidly changing technologies, social structures, methods of pedagogy and training, every institution must consider how to become and stay future-ready. Every leader must consider a new tool kit that will make the leader ready for the future. It has been my experience that impactful and appreciated leadership for a long time is characterized by the connection of aha occasions when the leader faces a steep challenge. Prime Minister Indira Gandhi is remembered for her responses when her party was split, Bangladesh happened, four successive droughts ravaged the economy, and so on.

1 Mindsets

Mindsets are relevant only when they become visible in actions. Bearing this fact in mind, it is worth noting the seven distinct mindsets of a future leader:

- Enthusiasm: high commitment and motivation to the cause.
- Obsession: The future will not be like the past
- Curiosity: about how the future leader will differ from the past.
- Acceptance: Embracing the future is more challenging than embracing the past.
- Realization: Leaders' style and content change, but not the core.
- Restlessness: to learn what is at the core of leadership.

R. Gopalakrishnan (✉)
Retired Director Tata Sons and Leadership Consultant, Mumbai, India
e-mail: rgopal@tata.com

I. Gratchev et al. (eds.), *Transformative Approaches to Career-Ready Education*,
https://doi.org/10.1007/978-981-96-7018-5_10

- Perceptiveness: to see the invisible, hear the inaudible, feel that which is not palpable.

The core of leadership in every domain is that after studying whatever can be taught, the future leader must keep learning what cannot be taught. As a professional rises, self-taught skills become more important than taught skills.

2 Practices

What practices should a future leader adopt to be effective? Five practices:

i. Listen more than speak
ii. Empathy to feel more than think
iii. Adapt while maintaining precious values.
iv. Reflect on experiences quietly
v. Innovate by practising the new things learnt.

3 Future-Ready Institutions

Long-living institutions—academic, corporate, scientific, and charity—must nurture values by developing sound, future-ready values and preserving those values across generations of leaders.

Although set in the context of India's freedom struggle, Gandhiji practised the philosophy of truth, ahimsa, and satyagraha. Symbols like a loin cloth, charkha, and the Dandi march iconized his practices. His narratives and rituals, such as group prayers, fasting, and jail terms were continual reminders to people.

Nurturing future-ready institutions accelerates a national economy because such institutions are enormous assets. The contributions over decades by century-old institutions like Ramakrishna Mission schools, Presidency College in Kolkata and Chennai, St Xavier's in Mumbai, and Indian Institute of Science in Bengaluru bear testimony to the observation. Future-ready institutions are designed to live long by preparing future generations, advancing the frontiers of knowledge, creating jobs, innovating technologies, treating people decently, and inspiring the community while being sustainable, humane, and enlightened.

Here is a catch: Setting up an institution with future-ready solid values is difficult. Remaining that way well after the founder's time is even more difficult.

There are models of excellence. Early leaders may implant deep-rooted and socially relevant values into the organization's DNA—the Indian Institute of Management Ahmedabad, Tata Institute of Fundamental Research Mumbai, and SP Jain Institute of Management and Research in Mumbai (SPJIMR), for example.

4 A Metaphor

Metaphorically, values flow like a long river from its origin until the river meets the sea.

From a trickle at its origin, the river has become bigger due to land gradients and hydrology and has become more impactful within society. Rivers bring bounty for the people whom they serve. As the river follows nature's contours, it may appear complex and intimidating. The Wide River develops its character, like ideas, and develops complexity due to icons and narratives. A vigorous delta rushes to meet the ocean in its final stretch and loses its identity.

This metaphor applies to the development of intellectual ideas and religions. Complexity gets added through narratives, icons, and rituals. That is the essence of all leadership. The delta represents the multiplicity of rituals to honour the same philosophy with different ways of expressing the simple idea.

This river metaphor provides a four-stage mental model that institutions must have: P-I-R-N, standing for Philosophy—Icons—Narratives—Rituals, just as it happens with the birth and spread of the world's greatest ideas in religion, science, and philosophy.

Institutions, too, must have a strategic intent, a well-thought-out and articulated philosophy characterized by simplicity. It can take them decades to reach their strategic intent. For example, SPJIMR wishes to "help practitioners with applying new theories of management knowledge." IISc states its strategic intent to "be among the world's foremost academic institutions through excellence in research ...." Finding out why many promising institutions failed to preserve values is noteworthy. There are examples of great academic institutions that have lost their way somewhere during their journey.

Although the following stories are corporate, consider two food companies that gained traction due to two different wars.

First, the Camelot chocolate company, Cadbury (Toole, 2019). John Cadbury, the son of an abstinent Quaker, added cocoa to his tea and coffee trading list in the 1800s. John Cadbury was as determined as his father to improve the world around him. In keeping with this philosophy, the family opened a chocolate township called Bournville, a bit like Port Sunlight of Unilever and Jamshedpur of Tata. Cadbury greatly benefited from the First World War when its products were regarded as nutritious and convenient for the fighting troops.

By 1965, the leadership passed to thirty-four-year-old family descendant Adrian Cadbury (famous for the Cadbury Committee on Corporate Governance). Cadbury engaged McKinsey Consultants to improve his company; the Americanization of British industrial thinking had begun. There followed a management restructuring and a sale of the confectionary business to American drinks company Schweppes, which was dominantly profit-oriented. Early founder John Cadbury's communitarian and humane philosophy was lost in the change in ownership.

Second, an iconic British brand, Bovril, is a nutritious beef extract product. Responding to a call for nutrition for a fighting British army in the Franco-Prussian

War, an innovative Scottish chemist, John Lawson Johnston, formulated a beef extract branded as Bovril, later joined by a vegetarian Marmite. To consumers, the brand blended meat, myth, and magic. By 1960, the descendants could no longer competently run the listed company. The founder's philosophy of "healthy, warm nutrition for workers" gave way to other philosophies as the business changed hands, first under investor James Goldsmith and under Unilever, which was bought in 2001. A tangential and light-hearted observation is that a descendant of the creator of beef extract married a Hindu girl!

Rivers end their journey for many reasons, and so it is with institutional philosophies. Changes in leadership, new leaders wanting to make a quick impact, and national ideologies that guide society are frequent causes of losing founding values. Preserving future-ready values across generations of leadership is difficult for other reasons as well, such as changing societal values and appointing 'wrong' successors.

5 Why Future-Ready Values Are Challenging to Maintain

During a speech in America in the 1890s, Swami Vivekananda stated that the West was groaning under the tyranny of too many Shylocks while the East was groaning under the tyranny of priests! The world needed, and still needs, institutions which are neither in the control of Shylocks nor priests.

It is relevant to note that several iconic research and teaching institutions in, for example, America, were founded on the capital accumulated by questionably ethical trades such as slavery, opium, and so on. An entrepreneur may not have noble values and a purpose other than making money or creating his or her place in history. If the founder has a purpose, it may not be sustainable, humane, or enlightened. The founder's good purpose may not be embedded into the enterprise, so it wanes over time. A future successor may deliberately depart from the original purpose. Some corporate examples follow.

Walter and Roy Disney set up Disney Studios in 1923, saying, "If you can dream it, you can do it." Their four-point motto was: think, believe, dream, and dare, and their purpose was to give wings to human imagination. What is happening in its centenary year (Jones, 2023)? The company now faces turmoil for several reasons. However, the view of a Disney successor is interesting. In 2022, Roy Disney's granddaughter and an independent filmmaker, Abigail Disney, produced a documentary titled *The American Dream and Other Fairy Tales.* Her underlying message was that too many Disney employees could not meet ends, while the disparities between executive and workers' pay widened unconscionably. "The company leadership have led the company brilliantly and deserve their bonuses. However, the CEO pay in 2018 was 1424 times the median pay of employees, and that is wild. Reward the people who make the magic," she argued.

Consider another iconic American brand, Starbucks, and its legendary founder, Howard Schultz. The company has been in a dispute with its workers who wish to form unions to negotiate better terms for themselves. Moreover, Starbucks will not hear of it (NYT, 2023).

The disruptions due to COVID had racked up record profits for the company but left many baristas exhausted and embittered. Starbucks did nothing illegal but stonewalled, did not respond, and chased down recalcitrant employees at troublesome stores. This state of affairs does not sit well for a company whose purpose is "to inspire and nurture the human spirit—one person, one cup, and one neighbourhood at a time." By the way, Howard Schultz became CEO twice after retirement in the last two decades after botched attempts at inducting successors.

Are these symptoms of the company having lost its values and moorings? We must wait and watch.

Consider the Rockefeller business, principally Standard Oil, which was humongous in the 1900s. John D Rockefeller was considered the wealthiest human being in his heyday. Emma Calve, the French operatic dramatic soprano, was his friend but also became an admirer and disciple of Swami Vivekananda during his sojourn in America. She persuaded Rockefeller to visit Swamiji. Emma Calve recorded the conversation, which has been published. Swamiji suggested to Rockefeller that even if he was a hundred times richer than the average American, he was unlikely to be a hundred times smarter. 'Could it be that you are a mere conduit for wealth?' posed Swamiji.

In 1912, fifteen years after he met with Swamiji, John Rockefeller endowed the Rockefeller Foundation. Today, while there is no Rockefeller business, seventy descendants own a piece of the Rockefeller Family Fund, which continues to do many wonderful things for society and people (Advaita Ashram, 1997).

6 Is a Future-Ready Institution an Idealistic Goal?

For sure, the concept is intellectually appealing but is so difficult to achieve that leaders begin to wonder whether they are practical. After all, there is a widespread view that most 'reputed' institutions are too often founded on skullduggery and ethical gore, that successful leaders have always been a bit egoistic and dictatorial rather than focusing on human relations or community, that effective leadership demands a heavy touch of autocracy and visible firmness.

Based on my professional experience, I dispute this view, even though I appreciate why such a view could exist. Scottish-American philosopher Alasdair MacIntyre dismissed the possibility that a manager in a for-profit business could ever be virtuous.

It is true that values-based institutions are not paragons of unblemished virtue; through their history, they can be faulted for aberrations, much as every great person exhibits human flaws. For the most part, institutions are virtuous through their long-term behaviour and actions.

They protect the resources that enable them to conduct their enterprise—*pancha bhuta* elements (earth, water, fire, air, and space), community, employees, vendors, and indeed all stakeholders. They are not driven only by money; they operate with fairness in terms of human relationships. To adapt a phrase, such enterprises practice Inayat (kindness), Ibadat (service), and Insaniyat (humaneness). They make the ruthless decisions their business demands but likely implement them compassionately.

They exhibit "spirituality" borrowed from HBS Professor Geoffrey Jones (Jones, 2023). He defines spirituality as "an implicit and explicit belief in the interconnectedness of all life and the planet that translates into a set of guiding principles and values." Interestingly, Swami Vivekananda said in the 1890s that the solution to the intense and fractious problems of the world at that time—indeed even today—was to "flood the world with spirituality." Note, spirituality, not religiosity!

Values-based institutions are giants of conscientiousness, though not necessarily giants of fame and size. These institutions practise exemplary behaviour towards their community. Conscientiousness giants may be big or small, well-known or unknown, but they are who they are.

An example of a big giant is The Cummins Engine Company, founded around the 1920s in Columbus, Indiana (Cruikshank & Sicilia, 1998).

Its explicit purpose was "to provide gainful employment to the youth of Southern Indiana." Clessie Cummins was a superb technician with a searing fascination for machines. Being an inveterate tinkerer, he sought jobs where he could earn while learning about his fascination, which, according to his account, "acted like a drug." He found a job as a chauffeur with a wealthy Irish-American financier, Mr Irwin, who helped finance the Cummins start-up. After the 'patient capital' had been stress-tested over about a decade, J Irwin Miller took charge of the company. He ran it for many years. The Cummins Engine Company took so much interest in their town of Columbus that upon his death, the *New York Times* observed in his obituary, "He and the company were instrumental in changing a decaying Columbus into a showcase for buildings." Well before companies were compelled to respond to calls for gender and racial diversity, Miller prompted his HR colleagues to hire African Americans for more senior positions.

Cummins operates in India too in two avatars: *first*, as a listed company called Cummins India, and *second*, as an unlisted joint venture with, who else but their brother-in-causes, Tata Motors!

That is why it is difficult to agree with distinguished philosopher Alasdair MacIntyre. Undoubtedly, institution-building is an idealistic goal, but it is practical. Everyone should strive for that goal.

7 The Power of Neeyat

Neeyat is not a novel idea or jargon. It has always existed. The word *neeyat* we use has its etymology in Arabic—the actual intent. The relationship between *Neeyat* and behaviour can be imagined to be similar to that of Gangotri to India, Vedanta to Hinduism, or the Preamble to the Indian Constitution. *Neeyat* is pure, inspiring, and uncluttered. However, it gets interpreted as development. Narratives, rituals, fables, and interpretations are added with distance and time; the Ganga no more accurately resembles the original Gangotri.

Likewise in institutions, strategists advise leaders to continually engage with the *neeyat or purpose* of the institution (the reason for its very being) and to ensure that its expression is faithful to the original intent. Apart from thinking about and appreciating the concept, there are twin practical leadership challenges: while *neeyat* is essential in the early stages, it also requires continual renewal and enhancement.

Jamsetji Tata stated that his company exists to serve the community, and Tata still does so. Jamsetji was surely influenced by the Zoroastrian principles of *Huma, Hukhta, and Hvarshta* (good thoughts, words, deeds).

William Hesketh Lever stated that his company exists "to clean the teeming millions all over the world". He diligently applied his Congregationalist principles in business. Unilever still cleans people. Andrew Carnegie (Carnegie Steel) was influenced by Swedenborg values, and Thomas Watson Sr (IBM) by Presbyterian ideas. John Cadbury was determined "to improve the world around" him and was driven by his Quaker principles.

In short, future-ready institutions demonstrate IQ, EQ, and AQ, which stand for Intelligence, Emotion, and Adaptability Quotients. These impart resilience to adapt to tectonic changes in technologies, society, and competition without ignoring transformation's human and relationship aspects (Gopalakrishnan & Bhattacharyya, 2024).

8 The Future-Ready Leader

It is worthwhile to reverse the traditional mental model of masculinity and power constructed through literature and books based on archetypal male leaders like Alexander of Macedonia, Chengiz Khan, Winston Churchill, Jack Welch, etc. These examples are from the fields of battle, politics, and enterprise. Of course, alternative models like Mahatma Gandhi, Nelson Mandela, Martin Luther King, and Vikram Sarabhai exist. In the future, influential leaders of the latter variety will be written about more than the leader's stern-eyed, tough-talking, square-jawed imagery.

Undoubtedly, India is at an exciting juncture, thanks to the palimpsest of policy and reform over several decades. These are times to avoid distractions. Among the critical challenges for Indian institutions is preparing 'The Tomorrow Leader' for the future society. What will be the impact of artificial intelligence and technologies on the future leader?".

9 Capability Building Among Leaders

Leadership exists in all domains—academic, political, public service, defence, enterprise (which happens to be my field of experience), and many others. Although one can slice and dice skills by domain, it is instructive to explore what has remained common from Alexander the Great to modern times, irrespective of timing or domain—the so-called eternal principles.

Successful leaders over the centuries have displayed three critical capabilities.

First, there is the capability to learn what can be taught, viz, stuff that is teachable. Alexander learned from Ptolemy and Aristotle. This is called *cognitive capability.*

Second, a curiosity and penchant for tracking changes in society and technology and distilling what is relevant to the immediate times. This is called *adaptive capability.*

Third, sensitivity and empathy for people relationships and the vagaries of human behaviour. This capability is called *behavioural capability.* The bouquet of these three capabilities influences an individual's leadership style and, more importantly, leadership effectiveness. This model is relevant and crucial, irrespective of the domain in which leadership is exercised.

The first capability, relating to teachability and domain-specific cognition, should not be generalized in a column like this. Defence people are trained in a Defence or Military Academy; enterprise people are trained at Institutes of Management, and academic people in teaching and research institutions.

During the first third of a career, irrespective of the domain, the demonstration of cognitive capability accelerates advancement. Bosses think well of young officers who learn quickly and apply their knowledge to benefit the institution.

In the second one-third of a career, it is essential to learn new skills without losing too much touch with the skills of the previous phase. As middle-level leaders position themselves for senior-level responsibilities, they must quickly adapt and incorporate adaptive and behavioural capabilities into their own leadership.

In the final one-third of the career, the mastery and deployment of behavioural capability is paramount. Then, we learn what every professional learns at that stage: "It is easier to predict the movement of the stars than to predict human behaviour." (paraphrasing an attribution to Sir Isaac Newton).

When we refer to people who are 'not with the times', we use age as a rough indicator. However, we mean their adaptive and/or behavioural capabilities have atrophied. Consider adaptive capability. A person who has been a print journalist par excellence may have missed the television or digital trends in the media environment. A defence person trained to think of workforce and armaments as the principal warfare tools might need to catch the new developments of remote armaments or cryptology.

Consider behavioural adaptation. One example is a manager or academic who might have risen impressively for demonstrating superior knowledge usable by subordinates. Such a rising star may need to prepare to learn from more knowledgeable youngsters later. Another example is the ability to 'let go.' After suitable training, junior people must be given the responsibility to discharge that responsibility and then be allowed to perform without close oversight. A boss must learn to let subordinates make mistakes—not ones that will sink the institution, but the self-learning blocks for the upcoming person. This is tough because the boss is still accountable to the broader stakeholders.

'The Tomorrow Leader' is vital when discussing a future-ready leader.

10 Losing Touch with Reality

Why do many leaders struggle hard and work diligently to rise? Once they reach the pinnacle, they falter. One reason is arrogance. Another is that they lose touch with the realities of the ground situation.

Why do the talented or top leaders tend to lose touch with reality? There are many possible reasons–all of them are plausible. Upcoming leaders tend to become bound by administrative routines and meetings. They are likely surrounded by people who wish to please them, so they whisper good things about the leaders. This inadvertently creates a bubble chamber in which the leaders tend to reside. Since they constantly receive positive signals about their leadership, they may become self-obsessed, and this results in selectively listening to positive news and inadvertent screening out of criticism.

In other words, the strength that catapulted them to the top becomes a weakness once they are there. Knowing that a top position can cause any great leader to lose touch with ground realities is an important leadership lesson.

Leaders are too often portrayed in writings and movies as forceful demagogues and 'leading from the front.' There is a new book authored by Ravi Kant, who retired as the Vice Chairman and CEO of Tata Motors after successfully leading the company through rather turbulent times. He is a low profile, quiet achiever, quite the contrarian compared to the loud, flashy leadership style of those portrayed on magazine covers. However, he was regarded as an effective leader at Hawkins, Titan, LML, and Tata Motors. He has co-authored a book that is in keeping with his personality (Kant et al., 2023). He writes about how leaders can keep their feet on the ground and their heads in the air.

The book sets out the principles of quiet leadership or leading from the back through the story of a bright leader, Shiv Kundra. He rises to the top amid great accomplishments and accolades from his company leaders. Suddenly, he falters. Why? He cannot let go. He micromanages his colleagues to the point of utterly frustrating them. The company directors are bewildered.

On the one hand, they value Shiv Kundra as an accomplished leader; conversely, he becomes his career enemy. Hence, they arrange some self-reflection through his exploring three questions: *First*, how should I be? *Second*, how should I deal with the team? *Third*, how should I deal with the task?

The second question is of direct relevance to the subject of this essay. Ravi Kant argues that you stay grounded by people relations–with trust, collaborating, and leveraging their strengths. That sounds pretty straightforward, but how does an upcoming leader practice these when the pressures of change, work schedules, and stiff targets make leadership life peripatetic and stressful? The book tells the reader a few things, but I recall an old Korean proverb, "You must slow down precisely when you have to reach faster."

The CEO Chairman of Hindustan Unilever would earmark 8–10 days every year for what used to be euphemistically called "Field Visit Day". On those days, the CEO would visit the market or the factory to interact, in my words, 'with the least powerful people in the organization': distributors, salesmen, factory workers and supervisors, for example. Moreover, that day would be spent without any intervening layers being present. In that way, the CEO would stumble upon ground realities through the 'truth-telling' of front-line employees and associates.

As the reader can see, this is no rocket science; it is the essence of simplicity. To recall Justice Oliver Wendell Holmes, "For the simplicity on this side of complexity, I wouldn't give you a fig. However, I would give you anything I have for the simplicity on the other side of complexity." Moreover, India's most simple Prime Minister since independence demonstrated this virtue.

In the early 1960s, Lal Bahadur Shastri had been told much about Dr Kurien's promising Operation Flood. He decided to visit Anand and see for himself but demanded that he do so incognito. This simple request became very complex with the protocol department, security department, and concerned ministries. Somehow, they solved the relevant problems—Shastriji stayed overnight in an unsuspecting farmer's house and discussed animal husbandry and milk economics. Imagine the farmer's shock when he later found out who his visitor was (Gopalakrishnan & Bhattacharyya, 2024)!

This field visit enabled focused government support for the Amul program. After fifty years, India can be proud that a milk-deficit nation has become a milk-surplus nation—thanks to grounded leadership and leading from the back on an important national priority.

11 Date Expiry of Leaders

It is helpful to ponder whether leaders make history or history makes leaders. What is undoubtedly correct is that history does not recognize leaders in the way that leaders thought they were making history. It takes work to get to the top. It is more challenging to stay on top. However, it is toughest to vacate the top. Leaders who experience scrambling up and staying on top sometimes need to improve on these toughest tests. A top leader was asked how he could be sure that he had selected the right top leader. He responded, "I do not know—not until the successor to the person I selected has performed." His response seems puzzling. Maybe one leader's performance is a flash, and maybe his successor is too, but if the third generation is performing, then the leadership planning system has probably been institutionalized. This anecdote symbolizes how long-term a view must be taken regarding leadership planning, renewal, and succession.

Recall that Chandragupta Maurya created the Mauryan empire, his son Bindusara strengthened it, and his son Asoka expanded it rapidly before adopting pacifist Buddhism. Asoka stayed thirty-seven years in his role; the decline and end of the Mauryan empire began with his death.

Sun King Louis XIV was emperor of France for seventy-two years. He indeed strengthened France but left it severely indebted. When he died, the empire was left in the hands of his five-year-old great-grandson, who became Louis XV. Royal authority steeply declined leading up to the French Revolution.

Ali Khamenei has served as the Supreme Leader of Iran since 1989, Yoweri Museveni has governed Uganda since 1986, and Hun Sen served as Cambodia's Prime Minister from 1985 to 1993 and again from 1998 to 2023. Examining the state of these nations as they transition from long-standing leadership will provide valuable insights.

When it comes to institutions, a tenure of ten years, at the most twenty years, is appropriate for one leader. There are exceptions, of course, but, for sure, after some time, the law of diminishing returns starts to take effect. The discipline of an institution having and implementing a clear retirement policy is an absolute must. Top leaders should also prepare to avoid being hangers-on in one capacity or another.

Systematic leadership change helps the company renew. Tata Consultancy, Tata Steel, and Titan have enough management bandwidth to allow orderly CEO changes every ten years or less. Further, it is true that such CEOs are hugely valuable at the time of retirement. Almost in all cases, such CEOs, upon retirement, can redirect their experience and judgement in new directions. This renews the company as well as the retiring leader.

Vacating the top spot requires humility. Poet Rabindranath Tagore wrote a relevant verse, though not about leaders' tenure. A translation of the Bengali original is, "Who will assume my tasks asked the majestic setting sun? Upon hearing this grave question, the world was speechless and silent. A mud wick lamp said quietly, 'My Lord, within my small capacity, I think I can help'."

The lesson is that today's leader was a mud wick lamp to an earlier leader; remember to nurture the mud wick lamps under your blaze.

12 Ten Invaluable Skills

i. How to disagree without becoming disagreeable.
ii. How to talk truth to power.
iii. How to unravel complexity through the simplicity of an open mind.
iv. How to get a degree of convergence from wildly divergent views.
v. How to admit that you do not understand.
vi. How to balance rationality with intuition or sixth sense.

vii. How to admit mistakes early and set about correcting those.
viii. How to function with action with two opposing ideas in the mind.
ix. How to develop self-awareness.
x. How to rock the boat without sinking it.

13 Summary

In closing, it is worthwhile to note the key features of future-ready leadership that apply in every domain. A professional must feel deeply committed to the cause, and this manifests as 'unputdownable enthusiasm.'

The leader's 'fuel and energy' must develop certain characteristics, which have been mentioned in the paper. The future will be different from the past, for sure, and it is a core responsibility of leadership to 'rock the boat without sinking it.' This requires combining two opposites—imagination (about the future) and experience (from the past).

References

Cruikshank, J. L., & Sicilia, D. B. (1998). Engine that could: Seventy-five years of values-driven change at Cummins Engine Company (No. HBSP-0068/XAB). Harvard Business School Publishing Corp.

Gopalakrishnan, R., & Bhattacharyya, H. (2024). *Embrace the future: The soft science of business transformation*. Bloomsbury Publishing.

Inside Starbucks' Dirty War Against Organized Labor, NYT, 21 July 2023.

Jones, G. (2023). *Deeply responsible business: A global history of values-driven leadership*. Harvard University Press.

Kant, R., Paul, H., & Reck, R. (2023). *Leading from the back to achieve the impossible*. Penguin India.

O'Toole, J. (2019). *The enlightened capitalists: Cautionary tales of business pioneers who tried to do well by doing good*. HarperCollins.

Virtue, A. (1981). University of Notre Dame Press. *Notre Dame*, 384–397.

Vivekananda, S. (1997). *The complete works of Swami Vivekananda* (Vol. 9).

Chapter 11
The Role of Universities in Solving Collective Action Dilemmas

Francis Kuriakose and Deepa Kylasam Iyer

Abstract Collective action dilemmas are those problems in which there is a conflict between individual and group interests. Norms add a layer of decision rule to collective action problems. The role of institutions in mediating collective action dilemmas is well known. This chapter explores the role of universities as institutions in mediating collective action problems, specifically climate adaptation. The role of a network of institutions that includes universities in adapting to climate change is taken as an illustration of a collective action dilemma. The study aims to examine how theory adaptation can be facilitated to this problem, yielding rich conceptualization. The chapter concludes that the first two pathways of direct and indirect actions embody the role of universities through the value of impact. The other pathways of mediation and engagement demonstrate the role of universities through the value of influence. The study argues that universities successfully solve collective action dilemmas when balancing these contradictory norms well. Sustainable universities will incorporate these insights into their curriculum and practices and impact their students by making them more employable.

1 Introduction

Climate change adaptation is arguably the most significant problem confronting the planet in the near future. Manifest climate change emergencies are all around us in erratic weather patterns, rising sea levels, frequent natural disasters, and profound energy crises. Climate change threatens to roll back several advances we attained in human development, namely childhood health, life expectancy, food security, labour rights at work and opportunities for shared prosperity. Beyond human development, inter-species rights and what we owe our planet remain unavoidable questions. The

F. Kuriakose (✉)
Impact and Policy Research Institute, New Delhi, India
e-mail: francis.kuriakose@gmail.com

D. K. Iyer
School of Industrial and Labor Relations, Cornell University, Ithaca, NY, USA

I. Gratchev et al. (eds.), *Transformative Approaches to Career-Ready Education*,
https://doi.org/10.1007/978-981-96-7018-5_11

two main approaches to collectively confront climate change issues are technological responses that help us adapt to the consequences of climate change and institutional maneuvers that can coordinate such responses. The responses to climate change adaptation call for us to examine both technological solutions and institutional coordination at various scales and contexts to identify designs of structures and processes that work. That is why the United Nations defined climate change as "adjustment in natural or human systems in response to actual or expected climatic stimuli or their effects, which moderates harm or exploits beneficial opportunities" through the Paris Agreement on Climate Change in 2015 (UNFCCC, 2015). Adaptation is one of the two central approaches that is globally given focus and attention.

The role of universities in climate change adaptation is intuitively evident. As centres of advanced research and development, universities are one of the major sites of scientific, technological, and normative knowledge production. Universities and colleges are also large institutions that can introduce internal sustainability practices. While research has examined the role of campus practices in several national contexts (Filho et al., 2018, 2020), the main drawback of this approach is that it detracts our attention from other influential roles universities can enforce (McCowan, 2022). This is the first "research gap" in the literature we address in this chapter.

The other point of interest is that extant literature on universities' role in climate change adaptation is grounded squarely on empirical considerations. While a first-hand account of practices on the ground and their monitoring and evaluation is valuable for practice, it misses the larger perspective on more abstract points of comparison and inspection that would connect patterns of institutional design and processes and identify pain points and good practices. The big-picture abstraction of why an institution works or not is equally important when dealing with a global public good ripe with collective action problems, among others. It is to this paucity of theorization that we suggest a framework that could be the beginning of fruitful and productive Meso-level theorization for a fundamentally empirical field.

Universities are fast becoming central actors in many arenas of action that exceed their traditional teaching and learning roles. The National Education Policy (Government of India, 2020) that the Indian government has ushered in places a new and dynamic role for universities in national development and global cooperation by focusing on research excellence, collaboration, and skill development. It pegs the internationalization of education as one of the key drivers of knowledge creation, transfer, and dissemination. This illustrates how much education with an international dimension is vetted and heavily invested by ambitious national governments. In a world that faces the double whammy of climate change and technological reorganization of the workplace, all problems are at once global and local, with a substantial risk of collective action dilemma looming in the background. It is unsurprising to see how institutions of higher learning are becoming the foci of design, planning and implementation of necessary policies in a manner unprecedented in history.

In this chapter, we use a meta-analysis of published works that examine university's role in climate change adaptation and how sustainable universities can be vehicles for curriculum, practices, and employability hinged around these principles.

With this data, we argue that research needs to adequately engage with the institutional role of universities beyond internal sustainability practices. We call for a shift in the focus from individual institutions to a network of institutions that involve the universities so that a diverse array of roles can be intellectually analyzed. We also point out the predominance of an empirical bias in existing literature and call for rigorous and expansive conceptualization of institutional role in climate change adaptation. We extend the theoretical adaptation of the collective action framework to this research agenda concerning institutional climate change adaptation and demonstrate how doing so will help with problem identification and pathways of institutional interaction. We also illuminate how this approach benefits interdisciplinary research on public good provision in other areas, such as technological change.

2 Literature Review

For this study, we reviewed and engaged with three separate strands of literature. First, we turned to collective action theory, which we proposed to use as a theoretical framework in our study. Second, we examined the existing research on universities and collective action problems to understand the state of current research and identify potential strands for making our case. Third, we analyzed literature that made a business case for sustainable universities in relation to their role in career-ready education and employment.

2.1 Collective Action Problems

Olson's (1989) assertion that when individuals in a group share common interests, it does not automatically guarantee their cooperation led to the beginning of a collective action framework. Individual actions in collective interest are a form of public good that markets will not spontaneously provide following the law of supply and demand. Olson also reasoned that they must be provided by an intermediary institution such as a government or non-governmental organization because such groups are not exclusive to specific individuals but go to the entire group in question (Olson, 2012). There were two further conditions that Olson set forth for collective action to occur productively—that the number of individuals involved should be small, and they should have recourse to a selection of incentives. Collective action dynamics play out in providing public goods such as law and order, defence, and pollution mitigation. By formulating a collective theory of action, Olson rejected the belief that individuals would act in unison when they collectively benefitted from an occurrence. He argued that self-interested individuals would not contribute to such efforts in what he termed a "zero contribution thesis" (Olson, 2012). Collective action problems were theorized under various social dilemmas, such as prisoner's dilemmas. Some examples of collective action dilemmas are given in Table 1.

Table 1 Examples of collective action problems

Example	Collective action problem	Potential solution
Voting	A large number of votes is necessary to change the course of policy. If people are discouraged from exercising their votes, this could lead to political apathy	A community's belief in political efficacy can counter apathy and lead to higher voter turnout
Reducing carbon emission	Most individuals have to reduce carbon emissions simultaneously to make a difference. The cost of individual strategy is high for the return it gives that individual	Government regulation of activities can reduce carbon emissions
Switching to renewable energy	An individual switching to renewable energy is costly for the benefits it gives them	Incentive schemes that reward switching to renewable energy can be effective

Researchers like Heckathorn (1989) parsed collective action into first and second-order problems. He delineated collective action problems as confronting the problems of cooperation that Olson had described as a first-order problem. Heckathorn reasoned that the ability of participants to modify behavioural expectations through social norms would lead to a host of second-order problems. For instance, when collective action arises through group norms, each participant faces four options—full cooperation, complete defection, and two forms of conditional cooperation: private and hypocritical cooperation. He also argued that second-order cooperation was more fragile than first-order cooperation because of the sanctioning force of meta-norms, which would lead to a third-order collective action problem and so on.

Ostrom (2000), through her extensive fieldwork in developmental problems around the world, demonstrated various instances where the variation of individual propensity to reciprocate was evident. She thought through the problem of designing institutions that would facilitate conditional cooperation. For example, her research showed that when participants believed others would reciprocate, played the iterations in the long run and had face-to-face communications, they tended to work towards collective benefits. Furthermore, she argued that the rules of the game, such as competition among participants, frequent in-person communication and the ability to sanction benefits and penalties, improved overall cooperation. One of the points she emphasized deserved further exploration was how various social norms affected how the game was played out in several contexts.

In subsequent research, Feiock (2013) examined an "institutional collective action framework" where the problem of collective action was extended to institutions. His main argument was that due to the partitioning of rules and responsibilities to various actors, institutions face the problem of cooperation and defection. The mechanisms for overcoming these dilemmas differed in scope and enforcement depending on the institutional contexts. The solutions were based on collaboration risk weighed against cost considerations where the least cost and least risk option was preferred. Feiock's work not only extended the collective action framework to institutions but also brought the social embeddedness of institutions to the forefront.

In this chapter, we use the framework of collective action problems to understand how universities deal with climate change adaptation. We suggest how the current research direction can be recast to bring an institutional angle to the problem with its network embeddedness. We also examine the climate change adaptation problem as one of the first and second orders. By doing so, we hope future research demonstrates the centrality of universities in a network of institutions responding to climate change and the design of institutional networks that would improve cooperation over defection in climate adaptation.

2.2 *Universities and Climate Adaptation*

Owen et al. (2013) demonstrated that the vast majority of literature on climate change adaptation focuses on institutions larger than universities, such as countries. However, within universities, the major focus of climate change responses was centred around energy, the built environment, and transportation. The main framework was to understand climate change response as a measure of risk against vulnerabilities, with institutions facing the most vulnerabilities to attain priority.

We found that a set of studies illuminated universities' diverse roles in climate change adaptation. For example, Filho (2010) approached an examination of a university's climate change commitment through a global survey that probed awareness of these issues among college students in 43 countries. The survey revealed the regional variation in university's engagement in climate change in their pedagogy and the emergence of the internet as the source of further education.

In another instance, McCowan (2020) situated universities as central players in negotiating climate change mitigation and adaptation. He charted out the diverse roles that universities played and the pathways through which universities acted, which serves as an analytical framework for understanding the institutional role of universities.

Furthermore, Fazey et al. (2021) argued that universities must confront manifest, conceptual, and existential emergencies simultaneously to successfully manage the climate change crisis. Taking a systems perspective, they spelled out the infrastructure, capacities, management tools, and orientations needed to move to a transformational role that universities can occupy.

A second set of studies examined universities in their empirical settings. McCowan et al. (2021) compared universities in seven countries engaged in various types of climate change responses. They argued for a differentiating institutional system, protected finance, leadership, frameworks and networks, and institutional diversity as factors that might improve performance across the board.

Latter and Capstick (2021) examined 32 universities in the United Kingdom that have issued climate change declarations to see what these practices implied. Using critical discourse analysis, they argued that while these practices expedite universities' internal sustainability orientation, they also detract them from taking more substantive actions externally.

The problem of measuring roles and responsibilities was of research concern. McCowan (2022) analyzed whether university responses to climate change could be meaningfully measured. He proposed a measurement framework that involved the dimensions of education, knowledge production, and public engagement.

While examining extant literature on universities' role in climate change responses, it was found that only a little research happened in this field. Furthermore, most research attempted to introduce a framework to categorize empirical findings. The paucity of active theorization is apparent in this field. Additionally, most of the studies on climate change demonstrated the focus on internal sustainability programs, revealing a limited imagination of the conception of the university as a stakeholder and embeddedness with other institutions. Our chapter stems from these gaps in the literature, and we sketch out how an argument for theory adaptation can encourage new strands of research moving forward.

2.3 Sustainable Universities and Career-Ready Education

The third strand of literature examines how sustainable universities can impact career-ready education. The business case for sustainable universities calls for embedding sustainable practices within the curriculum and policies of the university and tying it with the students' employability. Patrick et al. (2008) examine American case studies of universities' green practices to highlight the role of education and outreach, including expanding the curriculum, selecting and training eco-reps, greening Greek life and having sustainable practices associated with laboratories and other high-energy interventions within the campus.

By examining the case of sustainable universities, several studies argue that they can impact education trajectories by incorporating sustainability competencies and practices in curriculum, extracurricular activities, and community partnerships (Levesque, 2017; Purcell et al., 2019). Delving into specific programs, Kumar and Kim (2017) discuss the development of sustainability offices, recycling and waste reduction programs and certification as specific measures that campuses can design and implement to be sustainable institutions.

More recently, studies have assessed sustainable universities using the three pillars of curriculum, practices and investment behaviours (Nikolaou et al., 2023). These studies have sought to remedy the barriers to adopting sustainable practices that earlier literature had flagged off, including overcrowded curriculum, perceived irrelevance by academic staff and limited awareness, expertise, institutional drive and commitment (Dawe et al., 2005). Embedded sustainability is a concept gaining ground as a way to make sustainable practices resonate as a business case within the university (Cote, 2021; Visser, 2023; Whelan & Fink, 2016).

One of the main pieces of evidence in this strand of literature is the impact of sustainability on the employability of its students (Chen & Gan, 2021). Researchers also call for diversifying assessment tools to be genuinely impactful within the universities and beyond (Cohen, 2022; Dawodu et al., 2022).

3 Method

We discuss the research design through the data collection process and a description of the data. Data collection occurred through a two-step process, which is described first.

3.1 Data Collection

3.1.1 Exploratory Search

First, we conducted an exploratory investigation of literature about universities engaging in climate change-related work. We went through a prominent scholarly search engine and applied a broad range of keywords related to universities, climate change and collective action problems from 2008 to 2024. We excluded reports because they are a non-theoretical output of research. The search yielded close to 100 results. In light of the United Nations' call for climate change adaptation as the priority of action and a move away from climate mitigation in 2015, we decided to search for publications between 2015 and 2024. Concurrently, we also designed a search protocol for a systematic review that included a literature search strategy, coding scheme and inclusion criteria.

3.1.2 Systematic Search

As the first step, we devised a set of search terms that included combinations of "university," "research institute," "climate change adaptation," and "collective action problems." The period of analysis was publication between 2015 and 2024. For each set of keywords, we searched for links to publicly available articles. On average, a search through the first ten pages yielded the most relevant results.

At the first level of filtering, we included only publications in English. The types of publications included were peer-reviewed publications such as research articles, book chapters, edited books and dissertations, and patents, teaching notes, reports, and manuals were excluded. For publications with multiple versions, the most recent version was selected for inclusion in the database.

At the second level of filtering, we used accessibility and thematic boundary conditions. Articles behind a paywall were accessed through our university-subscribed accounts. Those beyond the subscription list and requiring individual purchasing were dropped. Similarly, a few articles with corrupted links were omitted from the study. In the content, we left out those studies that merely recommended that the university's role should be explored further. This resulted in a final sample of 30 publications. The data collection method is described in Fig. 1.

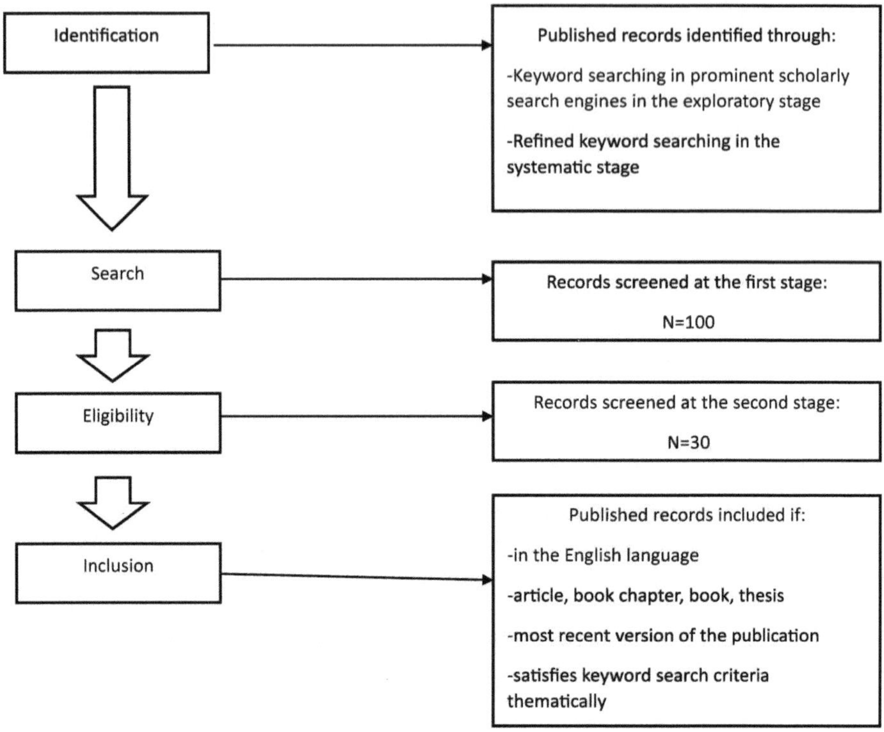

Fig. 1 Systematic search

3.2 Data Description

We used a database of 30 publications. A bibliometric review of climate change research between 1980 and 2014 indicates that climate change adaptation literature is a relatively small but growing subfield compared with others, and social sciences are lagging sciences in research output (Haunschild et al., 2016). Given this context, the descriptive statistics illustrated by our sample should be interpreted accordingly. Table 2 gives an overview of the variables we collected for our descriptive statistics.

4 Findings

Through this chapter, we identify novel dimensions of an established construct of climate change adaptation by introducing a new theoretical lens of the collective action problem. We take the university's role as the institutional illustration to

Table 2 Variables and definitions

N	Variable	Definition
1	Publication type	Whether the publication is a journal article, book chapter or book
2	Sub-field of research	The publication falls into the sub-fields of science and social science
3	Focus of research	The central issue that the research focuses on is universities' role in climate change. We have eight foci of research[a]
4	Research design	Whether the publication is conceptual or empirical
5	Research method	The social science research method(s) used to answer the research question
6	Country focus	The country that is the site of research

Note [a]The eight research foci are year, publication type, field, issues, concept, research design, method and country focus

make the claim. From our data on the university's role in climate change adaptation, it is clear that overall, the number of publications on this research question is exceptionally low.

Based on this evidence, we discuss a few important trajectories that could shift this trend. At the empirical level, universities need to be emphasized as the lead stakeholders in climate change adaptation. Shifting the central focus this way would illuminate the centrality of relationships among various stakeholders in the knowledge transfer domain. Thus, we transform the policy-centred focus on monitoring and evaluating adaptation strategies to the actors involved in the policy process while maintaining the central construct of adaptation intact.

Several iterations of these empirical analyses encourage a conceptualization of the institutional role in climate change adaptation. We illuminate one such conceptual framework of collective action problem widely used in public economics as the lens that yields powerful insights for theory and practice. This framework is particularly apposite given the shifting focus of institutional interaction from one actor to a network of actors that would need cooperation and coordination. Recasting institutional adaptation as a collective action problem would encourage the market and the state to seek incentives and policies for cost-efficient coordination. The reasoning from empirical focus to theoretical shift is a form of theory adaptation where new dimensions of a construct become visible when treated through a new theoretical lens. Our schema of theoretical adaptation is depicted in Fig. 2.

Fig. 2 Schema of theory adaptation

4.1 Problem Identification

Next, we parse how conceptualizing knowledge transfer coordination as a collective action problem yields insights by identifying new, otherwise invisible problems. Problem identification is how levels and types of collective action problems are identified when institutions collaborate on climate change adaptation. Pathway identification is the process of mapping pathways of cooperation and stakeholder partners based on values that matter to institutions. Pathway identification is a method of solving the defection problems that problem identification throws up.

From the overall mapping of the relationship between empirical evidence and theoretical framework depicted in Fig. 2, we parse out the transformation steps. First, we argue that the theoretical framework primarily helps in problem identification. For example, collective action theory identifies first-order and second-order problems in the context of group behaviour. The first-order problem pertains to the original collective action dilemmas, such as freeriding and creating the tragedy of the commons. Freeriding is an instance in which an individual benefits without contributing to collective action. The tragedy of the commons is when individuals

Table 3 First-order problem examples

Problem	Definition	Example	Solution
Freeriding	Individual benefits without making a contribution	Streetlights	Levy is a tax that helps install and maintain street lights without individuals paying for their use
Tragedy of commons	Individuals deplete a public resource in the long run	Overhunting and overfishing	Regulation of hunting and fishing periods

use a public resource or commons for their interest and deplete it in the long run. Table 3 describes freeriding and the tragedy of commons with examples.

The second-order collective action problem arises due to the first-order problem through intra-group sanctioning norms. Since flouting cooperation induces group sanctioning systems, each institution has a four-fold choice of action: full cooperation, complete defection, hypocritical cooperation, and private cooperation (Heckathorn, 1989).

Underlying the sanctioning system or the second-order group norm effect is the presence of values. This implies that second-order cooperation, such as hypocritical cooperation, is more robust than private cooperation. Thus, the four-fold choice of producing a type of cooperation or defection at individual and normative levels explains the phenomenon of climate adaptation by making the decision-making structures and choices more apparent. The transparency of choices encourages us to think about conditions in which some choices become more prevalent than others and to design circumstances that channel decisions that optimize public good production. Figure 3 summarizes the problem identification described here.

4.2 Pathway Identification

We argue that using the two-step analysis of collective action theory on universities regarding climate change adaptation makes the action pathways explicit. This is the contribution of the conceptual framework to practice. As Fig. 4 demonstrates, identifying problems makes the pathways obvious through the values that institutions hold. For instance, the value trade-off between impact and influence is made explicit at the foundation. Impact and influence are both concerned with creating changes. While impact refers to the direct mode of causing transformation and is often tangible, influence is more indirect and intangible. The value of impact guides action that can be either direct or indirect. Direct action involves core activities that universities are concerned about, such as teaching and research. Indirect action may come about in the same teaching and research domain through external partners that can collaborate or co-create knowledge in some ways.

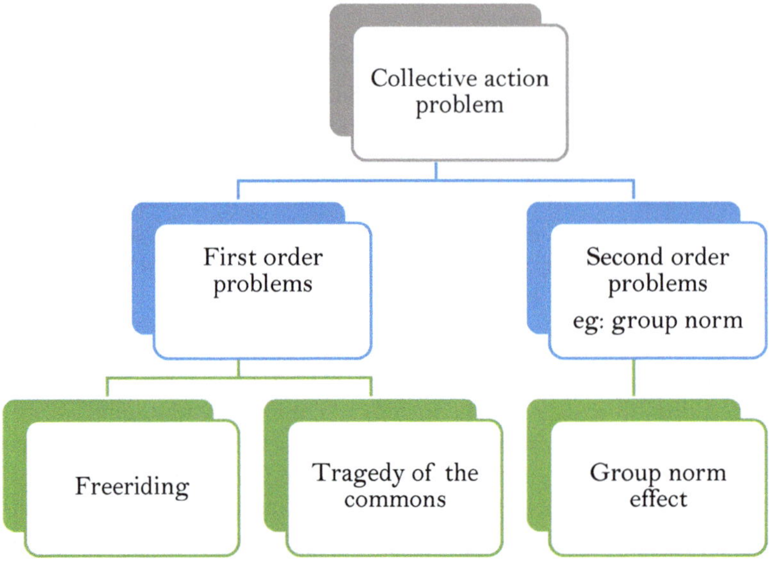

Fig. 3 Problem identification

Similarly, influence can come about through the mediation of universities as intermediaries in a network or through public intellectualism that directly engages with various publics. Furthermore, the pathways identify organizations that can engage in each type of activity. Thus, the theory helps in charting pathways illuminating the underlying values and their trade-offs, ways of achieving those values, and fruitful partnerships in attaining them. It is relevant to compare that existing research that does not theorize these values do identify some, such as teaching, research and mediation, but not all activities of interest and fail to see how they are interrelated. This, in turn, limits how much they can point out pathways of actions that are meaningful for the institutions.

5 Discussion

5.1 Theoretical Contributions

Our findings provide the perspectives for a new understanding of the institutional role in climate change adaptation. We take the illustration of the university as an institution to make the case. Our first argument is that shifting the focus from empirical cases of single institutions to nodal institutions renders the role of the network visible. For instance, from an exclusive focus on another type of not-for-profits as intermediaries in climate change adaptation literature, the examination of a nodal institution such

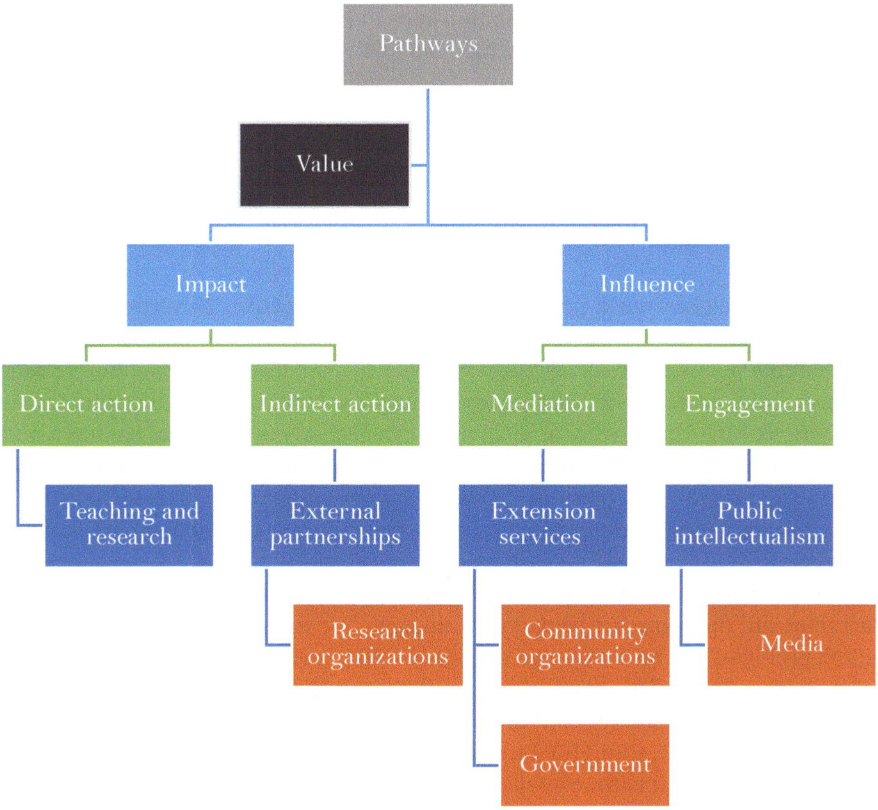

Fig. 4 Pathways

as a university brings forth the forward linkages with ground-level partners like civil society organizations and beneficiary groups and backward linkages to the industry and government that supports research.

Our second argument is to theorize the empirical cases from which patterns of generalizations can be meaningfully made. We make the case with collective action problem theory and demonstrate how using the theoretical constructs has implications for identifying problems of cooperation and defection among the network of institutions. We trace the first-order problem of cooperation against freeriding and the tragedy of commons. We also argue that solving the first-order problem through the sanction mechanism led to secondary problems of the group norm effect that encourages superficial cooperation more than robust forms of cooperation.

Finally, we trace the possible interaction pathways from identifying problems. The values undergirding the cooperation pathways are explicit at the first level of pathway mapping. Then, different options of cooperation and interaction are charted out. There are four preliminary pathways of direct action: indirect action, mediation, and public intellectualism to solve some of the collective action problems. They also identify possible partners and the specific relationships that warrant a stable equilibrium. We also note that the pathways can interact and evolve into interrelated paths. Further, theorization can delineate the primary and secondary paths and possible interaction channels among the various pathways.

5.2 Policy Implications

Our article sketches out a primarily theoretical perspective. Nevertheless, understanding the institutions with respect to climate adaptation in this manner has several policy implications. First, identifying a network of institutions and mapping potential partners shifts the focus of planning and budget allocation through the entire gamut of actors, not the siloed view of a single organization. For example, having a node and network of actors in mind might influence how teaching and research are done at the design stage by incorporating other actors as participants. This understanding might call for a modification in how extension activities are carried out in relation to a project.

The problem of cooperation and defection makes it explicit that incentives and sanctions are the route for behavioural nudges. For instance, first-order collective action problems can be confronted by incentivizing a sense of ownership and investment in the project. This entails having climate change adaptation practices in the institutions and researching and disseminating knowledge about them. The group norm effect has implications for the size of institutions involved. For example, if large institutions are paired with several small organizations, there should be guardrails to check that disproportionate sanctions are not placed on those who cannot bear them.

This, in turn, impacts the way feedback mechanisms like monitoring and evaluation will play out. Sustainable universities are organizations where insights are embedded in the design of the curriculum and practices. This embeddedness makes sustainability an organic feature that can be assessed, measured, and improved and impacts the employability of its students.

5.3 Limitations

The study depends on a dataset collected by a single method of limited keyword search on a digital scholarly platform. Forward searches on journals and other databases and collecting unpublished academic work can contribute to the richness of the data and help navigate the questions and framework. Without such triangulated data, the scope of interpretation is limited.

The second issue is that we have limited our focus to climate change adaptation to keep a tab on the sources and analysis of collective action concerning that phenomenon. Climate change responses are complex, interrelated strategies that include mitigation, adaptation, and other risk and vulnerability calculations. Due to the UN's emphasis on climate change adaptation as the focus since 2015, we decided this framework was pertinent to climate change response over other valid considerations.

Finally, the sample of countries and contexts we used in our study might over-sample countries and contexts already facing drastic climate change issues over others. Up to an extent, this trend is unavoidable in cross-sectional samples such as ours, but we argue that since we do not quantitatively measure metrics, the overall arguments vary very little.

6 Conclusion

A prominent feature of contemporary climate change adaptation research is the role of institutions, especially the not-for-profit sector. Through a systemic review and meta-analysis of extant literature, we demonstrated that university as a type of not-for-profit institution engaged in climate change adaptation is sparse. Using collective action theory, we illustrated how theorizing has implications for problem identification and solving the problem through pathway tracing. Furthermore, we also showed that shifting research focus from single institutions to a network of institutions leads to perceiving the relationship between different structures and processes that make up climate adaptation.

Our study is an attempt at theory adaptation. Theory adaptation is a rigorous method of extending conceptual understanding similar to theory development, typology and modelling. We used meta-analysis to make the case and used elements of theory from public goods literature to climate adaptation to argue for a shift in focus of analysis with consequences for theory and practice. This mix of literature review and analysis with a conceptual framework from another discipline is a method that can be adopted for theory adaptation in other fields.

We strongly encourage future researchers to employ various variables in their meta-analysis and explore contexts where such research designs are productive. These include understudied and unconventional settings that approach different scales of analysis, from small grassroots units to meso-level institutions and prominent macro factors. We hope that our systemic literature review, meta-analysis, and attempt at novel theoretical adaptation encourage a generation of scholarship on the nature and consequences of collective action. We believe such a research agenda can reset our current dialogues on reparation through a rational framework of cost and benefits. This might encourage more transparency and trustworthy institutional relationships to flourish.

References

Chen, Y., & Gan, N. (2021). Sustainable development of employability of university students based on participation in the internship promotion programme of Zhejiang province. *Sustainability, 13*(23).

Cohen, S. (2022). *The role of universities in the transition to environmental sustainability*. State of the Planet. Retrieved 19 April 2024 from https://news.climate.columbia.edu/2022/12/27/the-role-of-universities-in-the-transition-to-environmental-sustainability/

Cote, C. (2021). *Making the business case for sustainability*. Retrieved 19 April 2024 from https://online.hbs.edu/blog/post/business-case-for-sustainability

Dawe, G., Jucker, R., & Martin, S. (2005). *Sustainable development in higher education: Current practice and future developments*. Retrieved 19 April 2024 from https://www.sustainabilityexchange.ac.uk/files/sustdevinhefinalreport.pdf

Dawodu, A., Dai, H., Zou, T., Zhou, H., Lian, W., Oladejo, J., & Osebor, F. (2022). Campus sustainability research: Indicators and dimensions to consider for the design and assessment of a sustainable campus. *Heliyon, 8*(12).

Fazey, I., Hughes, C., Schäpke, N. A., Leicester, G., Eyre, L., Goldstein, B. E., Hodgson, A., Mason-Jones, A. J., Moser, S. C., Sharpe, B., & Reed, M. S. (2021). Renewing universities in our climate emergency: Stewarding system change and transformation. *Frontiers in Sustainability, 2*.

Feiock, R. C. (2013). The institutional collective action framework. *Policy Studies Journal, 41*(3), 397–425.

Filho, W. L. (2010). Climate change at universities: Results of a world survey. In W. Leal Filho (Ed.), *Universities and climate change: Introducing climate change to university* (pp. 1–19). Springer Berlin.

Filho, W., Frankenberger, F., Iglecias, P., & Mülfarth, R. C. K. (Eds.). (2018). *Towards green campus operations: Energy, climate and sustainable development initiatives at universities*. Springer.

Filho, W., Nagy, G. J., Borga, M., Muñoz, P. D. C., & Magnuszewski, A. (2020). *Climate change, hazards and adaptation options*. Springer International Publishing.

Government of India. (2020). *New Education Policy*. Retrieved 3 March 2024 from https://www.uil.unesco.org/en/articles/india-national-education-policy

Haunschild, R., Bornmann, L., & Marx, W. (2016). Climate change research in view of bibliometrics. *PloS one, 11*(7).

Heckathorn, D. D. (1989). Collective action and the second-order free-rider problem. *Rationality and Society, 1*(1), 78–100.

Kumar, A., & Kim, D. S. (Eds.). (2017). *Sustainability practice and education on university campuses and beyond*. Bentham Science Publishers.

Latter, B., & Capstick, S. (2021). Climate emergency: UK universities' declarations and their role in responding to climate change. *Frontiers in Sustainability, 2*.

Levesque, V. (2017). *Sustainability methods and perspectives*. Retrieved 19 April 2024 from https://pressbooks.pub/sustainabilitymethods/front-matter/titleandpreface/

McCowan, T. (2020). *The impact of universities on climate change: A theoretical framework* (Transforming Universities for a Changing Climate Working Paper Series, No. 1). Institute of Education, London: University College London.

McCowan, T. (2022). Can we measure Universities' impact on climate change? *International Higher Education, 111*, 17–19.

McCowan, T., Leal, W., & Brandli, L. (Eds.). (2021). *Universities facing climate change and sustainability*. Global University Leaders Council.

Nikolaou, I. I., Tsalis, T. A., Trevlopoulos, N. S., Mathea, A., Avlogiaris, G., & Vitalis, K. I. (2023). Exploring the sustainable reporting practices of universities in relation to the United Nations' 2030 Agenda for Sustainable Development. *Discover Sustainability, 4*(1), 46.

Olson, M. (1989). Collective action. In M. Milgate & P. Newman (Eds.), *The invisible hand* (pp. 61–69). Palgrave Macmillan.

Olson, M. (2012). The logic of collective action [1965]. In C. Calhoun, J. Gerteis, J. Moody, S. Pfaff & I. Virk (Eds.), *Contemporary sociological theory* (pp. 124–128). Wiley.
Ostrom, E. (2000). Collective action and the evolution of social norms. *Journal of Economic Perspectives, 14*(3), 137–158.
Owen, R., Fisher, E., & McKenzie, K. (2013). Beyond reduction: Climate change adaptation planning for universities and colleges. *International Journal of Sustainability in Higher Education, 14*(2), 146–159.
Patrick, D. L., Murray, T., & Bowles, I. A. (2008). *Campus sustainability best practices.* Retrieved 19 April 2024 from https://www.mass.gov/doc/campus-sustainability-best-practices-a-resource-for-colleges-and-universities/download
Purcell, W. M., Henriksen, H., & Spengler, J. D. (2019). Universities as the engine of transformational sustainability toward delivering the sustainable development goals: "Living labs" for sustainability. *International Journal of Sustainability in Higher Education, 20*(8), 1343–1357.
UNFCCC. (2015). *Adaptation and resilience.* Retrieved 4 March 2024 from https://unfccc.int/topics/adaptation-and-resilience/the-big-picture/introduction
Visser, W. (2023). *Business case for sustainability.* Retrieved 19 April 2024 from https://www.cisl.cam.ac.uk/news/blog/business-case-sustainability
Whelan, T., & Fink, C. (2016). The comprehensive business case for sustainability. *Harvard Business Review*, 21.

Chapter 12
University Education for the Future

Raghunath K. Shevgaonkar

Abstract University education is crucial for societal advancement and is closely linked to economic development. Understanding the evolution of university systems is essential for shaping their future direction. Initially, universities existed to generate, preserve, and share knowledge in an open and accessible environment while nurturing cognitive and critical thinking skills. However, the Industrial Revolution shifted the focus towards applying knowledge for societal progress and workforce development, leading to the widespread expansion of higher education. Universities primarily emphasize knowledge dissemination through teaching and research, with professors serving as the primary source of knowledge. The digital revolution dramatically transformed education by making knowledge universally accessible and redefining the role of professors as facilitators rather than sole knowledge providers. Technology-enabled asynchronous learning emerged, enabling self-paced education. The COVID-19 pandemic further underscored the effectiveness of technology-assisted learning. In this century, the pace of knowledge generation is rapid, challenging universities to design curricula that prepare students for an uncertain future. Higher education is expected to become more student-centric, allowing for flexible program design. Emerging technologies like AI, robotics, and biotechnology will redefine human needs and relevance. Sustainability is a pressing challenge, necessitating that universities prioritize education that fosters a sustainable and fulfilling society. Future universities must integrate modern techniques with ancient principles to promote ecological integration and ensure sustainability.

1 History of University System

University education is the backbone of any advanced society. There is a good correlation between the higher education system and a country's economic development. University systems worldwide have evolved over many centuries to reach the level

R. K. Shevgaonkar (✉)
IIT Bombay, Mumbai, India
e-mail: rkshevgaonkar@gmail.com

I. Gratchev et al. (eds.), *Transformative Approaches to Career-Ready Education*,
https://doi.org/10.1007/978-981-96-7018-5_12

that we see today. It is, therefore, worthwhile to look at the evolution of the university system while defining its possible future direction.

The university system's history is an engaging narrative that unfolds over centuries, revealing the evolution of institutions dedicated to higher learning and intellectual exploration. The university's origin can be glimpsed in the academies of ancient civilizations. India has a rich history of higher education dating back to ancient times. Universities like Takshashila were set up during the fourth century BC. Universities like Nalanda, Vikramshila, etc., which were learning centres and attracted scholars from different parts of the world, existed from the fourth to eleventh century AD. There were renowned Academies in Athens and the Library of Alexandria dating back to the fourth century BC, laying the groundwork for transmitting knowledge. These ancient universities played a significant role in intellectual and cultural history. They were not only centres for religious and philosophical studies but also places where various branches of knowledge, including mathematics, astronomy, medicine, and the arts, were taught and developed. However, it was in medieval Europe that the university system began to take shape as we recognize it today.

One of the earliest universities, the University of Bologna, was founded in Italy in 1088, and it stands as a testament to the enduring spirit of academia. This institution pioneered several vital principles, including academic freedom and self-governance. The curriculum was structured around a series of lectures, and students engaged in scholarly pursuits under the guidance of learned professors. The success of the University of Bologna set a precedent, and similar institutions began to emerge across Europe, with the University of Paris, Oxford, and Cambridge among the early adopters of this model.

As the Renaissance dawned, universities became hubs of humanist thought, fostering a revival of classical learning and encouraging the pursuit of knowledge for its own sake. The emphasis on liberal arts and critical thinking laid the foundation for a more enlightened and educated society. The seventeenth century witnessed the scientific revolution, a period of profound transformation in which universities became crucibles for empirical observation and experimentation, reshaping how humanity understood the natural world—the nineteenth and twentieth centuries marked a significant turning point in the university system's history. Higher education underwent a democratization process. The establishment of public universities broadened access to learning. During this period, they also shifted towards a research-oriented approach, with universities becoming critical players in scientific and technological advancements. The ivory tower expanded its walls to engage with real-world problems, and disciplines like medicine, engineering, and social sciences gained prominence.

Universities play an indispensable role in shaping global knowledge economies today. They serve as incubators for innovation, foster interdisciplinary collaboration, and address complex societal challenges.

2 Classical Objectives of Higher Education

The classical objectives of a university have evolved over centuries, reflecting the changing needs of society and the expanding role of higher education. While specific priorities may vary among institutions, particular core objectives have endured, shaping universities' fundamental purpose worldwide. Universities foster intellectual communities where scholars, students, and researchers engage in a dynamic exchange of ideas. This community creates an environment conducive to intellectual Growth, collaboration, and the free exchange of knowledge. University is a place where the truth is searched without any prejudice. It is where individuals can dive into their passion and explore their strengths.

A university's primary and most acknowledged objective is to provide a structured environment for teaching and learning. Universities offer degree programs across various disciplines, imparting knowledge and skills to students through lectures, seminars, and practical experiences. The emphasis is on academic rigour, logical thinking, and developing a well-rounded education. Through exposure to diverse perspectives and challenging academic content, students develop valuable analytical skills in their chosen professions and in navigating the world's complexities. Cultivating critical thinking skills is extremely important for identifying and solving complex problems. Through scientific inquiry, experimentation, and exploration, universities actively contribute to creating and advancing knowledge in various fields.

In addition to academic and professional skills, universities aim to contribute to the holistic development of individuals. Character development involves instilling values, ethics, and a sense of responsibility, preparing students for successful careers and active and responsible participation in society. Universities make their graduates capable of earning their livelihood.

3 Higher Education After Industrialization

In an industrialized world, universities play a multifaceted and pivotal role in driving progress, innovation, and societal development. Their functions extend beyond traditional academic pursuits to actively contribute to the economic, technological, and cultural advancements that characterize industrialized societies.

In the industrialized world, universities have become the epicentres of research and innovation, conducting studies that push the boundaries of knowledge. There is a thrust on applied research that has direct applications in various industries. Universities collaborate with businesses to develop cutting-edge technologies, improve processes, and contribute to industrial innovation. University research outcomes can lead to breakthroughs in science, engineering, medicine, and technology, enhancing industrial productivity and competitiveness. This symbiotic relationship between academia and industry fosters a dynamic ecosystem where ideas are translated into practical solutions. Universities contribute to practically applying research findings in the industrial sector through collaborations, licensing agreements, and spin-off companies. Start-ups and established companies benefit from the expertise and intellectual property generated within university research departments.

In an industrialized society, a skilled and educated workforce is crucial for economic Growth. Universities play a key role in preparing individuals for careers in diverse industries through education and training programs. They offer specialized courses and degrees tailored to the needs of the labour market, ensuring that graduates possess the skills and knowledge required for the industrial sector. This alignment between education and industry needs fosters a seamless transition for graduates into the workforce. By instilling a foundation of both technical expertise and versatile skills, universities empower individuals to contribute meaningfully to the industrial workforce.

The industrialized world demands a continuously evolving workforce. Universities offer opportunities for professionals to engage in lifelong learning, providing avenues for skill development, updating knowledge, and staying abreast with industry trends. Through executive education programs, workshops, and certifications, universities contribute to the ongoing professional development of individuals within industrial sectors. Continuing education programs also serve as a mechanism for industries to upskill their workforce in response to technological advancements and changing market demands.

Universities engage in research and scholarship that extends beyond technological advancements to address societal challenges. Universities are critical in debates impacting industry and society, advocating for responsible and sustainable practices. In an industrialized world, they shape public policy and regulatory frameworks that balance economic interests with social and environmental considerations.

It is then clear that universities' role in an industrialized world is expansive, encompassing research, education, innovation, and societal engagement. Their impact extends far beyond the campus, influencing the broader society's economic, technological, and cultural landscapes. As they continue adapting to emerging challenges and opportunities, universities remain indispensable partners in the evolution of industrialized societies.

4 Status of University Education Today

Today, university education has become an essential part of society. It involves community engagement, addressing societal challenges through research, and fostering a sense of social responsibility among students. Although the gross enrolment ratio in developing countries is low at this point, they are working on enhancing it by making university education more accessible to the lowest strata of society.

In recent years, one major change in education has been the digital transformation of higher education. The integration of technology in university education has changed the paradigm of teaching–learning. Digital tools, online platforms, and virtual classrooms have become increasingly prevalent. The global COVID-19 pandemic further accelerates the use of technology. The digitalization of education, while making it more accessible, has posed challenges related to the digital divide and the need for pedagogical adaptation.

Online resources have made access to information easy for everyone, irrespective of the learners' geographical location and economic and societal status. International collaboration, student exchange programs, and joint research initiatives have become common. In such a globalized education, universities are putting extra effort into creating inclusive environments that celebrate diversity in terms of student demographics, faculty representation, and curriculum content.

Today, the main thrust of university education is the employability of graduates. There is a growing emphasis on aligning university education with the job market's needs. Employability and the development of practical skills are vital considerations. Some universities are integrating internships, industry partnerships, and practical training into their programs to bridge the perceived gap between academia and the workforce.

Although imparting knowledge to students is an important aspect of a university, today, universities are ranked more based on their research contributions. The pursuit of cutting-edge knowledge, scientific breakthroughs, and technological advancements is taken as the central mission. Research output, collaborations with industry, and the ability to attract top-tier faculty and students are critical indicators of a university's standing.

While the university system has expanded manifoldly recently, there have been many challenges before universities.

One of the significant challenges that the university system faces today is to make education more affordable. The cost of higher education, including tuition fees, housing, and textbooks, has continued to rise in many countries. This has led to concerns about accessibility, as students may face financial barriers to pursuing a university education. Student debt levels, particularly in some countries, have become a significant societal issue.

Another challenge that universities face today is the students' less inclination towards conventional degrees. This is primarily due to an excessive thrust on employment-centric education. The demand for non-traditional credentials, such as micro-credentials, certificates, and online courses, is rising. Lifelong learning and the need for continuous skill development are reshaping the traditional notion of a university degree. Universities are adapting by offering flexible learning options and recognizing a broader range of educational achievements.

The challenge posed due to easy access to information is on two fronts: (i) Standardization of education, cultural inclusivity, digital divide, and impact on the functioning of institutions. (ii) Assessment of the students. Traditional methods of assessing student performance may only partially capture a student's capabilities and, therefore, need re-evaluation. Educators are exploring more dynamic and inclusive assessment methods like project-based assessments, portfolios, and competency-based evaluations.

In summary, while universities continue to be pillars of knowledge creation and dissemination, they face complex challenges related to access, affordability, relevance, and societal impact. The current university system, therefore, needs a change. To keep pace with technology-dominated societal changes, the university system needs to dynamically evaluate society's future needs. It is, therefore, important to assess the changes that society is likely to undergo in the future. This requires a 'Future Envisioning Exercise'.

5 Envisioning the Future for Developing Future Universities

Envisioning the future for the next 50 years involves exploring a range of scenarios that consider technological advancements, societal changes, environmental shifts, and geopolitical developments. Today, we live simultaneously in two worlds: the physical world and the Cyber world. A large fraction of our activities are governed by the cyber world. While the physical world is reasonably predictable, the cyber world is mostly unpredictable beyond a decade or so. The future, therefore, is inherently uncertain, and multiple factors will shape the trajectory of our world.

At this point, it is worthwhile to ask, 'Is 'envisioning the future' a forecasting exercise or a path-defining exercise'? The answer to this question will decide the direction of the future universities. If we take the envisioning exercise as a forecasting exercise, one can make some projections for the technologies of the future and their impact on society. While forecasting, it is assumed that the current development model continues, and society keeps adapting to new technologies. The technologies evolve organically with a push–pull arrangement with society.

In this scenario, rapid advancements in technology will decide the future. Some technologies (as we see today), like artificial intelligence, quantum computing, and biotechnology, will contribute to shaping human lives. However, technology is a double-edged sword. If used correctly, it can create a utopian world. However, the same technology, if not directed properly, may create a technology dystopia. A future might emerge where technological advancements lead to unsustainable consumption of natural resources, social inequality, increased surveillance, and loss of privacy.

In the forecasting approach, since the current development model is extrapolated, universities remain silent contributors to the market's needs. Their primary job is to train the workforce as per industrial requirements. The data presented at various international forums amply suggests that the current development model is not very inclusive. Wealth may get concentrated in a few pockets, and the social and economic divide may continue to rise. The model may increase the disparity in the world, making it socially unstable. The model also puts tremendous pressure on the natural resources. At this point, it is difficult to assess how long this development model will sustain!

The current development model does not lead to a sustainable society in kind and spirit. The future development model needs new thinking and a non-conventional approach. The world needs not to evolve organically but to progress in a well-thought-out direction. Therefore, envisioning the future should be a 'path-defining' exercise and not merely a 'forecasting exercise'.

In the directed model of evolution, the physical and mental health of society and its sustainability will be the prime objectives. There will be a correction to the existing development model that has evolved over the last century. This scenario would focus on renewable energy dominance and natural resource management. Advances in solar, wind, and other renewable technologies will lead to their widespread adoption, reducing reliance on fossil fuels. This transition will mitigate the environmental issues and reshape geopolitics by altering energy dependencies and reducing conflicts over oil resources.

A cultural renaissance is necessary for a happy society. Therefore, the 'path-defining' scenario should focus on creating a future with a renewed emphasis on the arts, literature, and cultural exchange. This scenario highlights the importance of preserving and celebrating cultural diversity as a means of promoting empathy and global harmony. In this approach, geopolitical power dynamics may undergo significant shifts. The rise of new global players, changes in alliances, and reconfiguring economic and political blocs could reshape the international order.

Envisioning the future requires a holistic understanding of the complex interplay between technological, social, economic, and environmental factors. It is a dynamic process that involves anticipating possibilities, preparing for uncertainties, and actively working towards a future that aligns with our shared values and aspirations. The path ahead will be influenced by our choices today, the policies we implement, and our collective ability to address global sustainability challenges.

6 Impact of Technology on Future Society

The last few decades have seen a catastrophic growth of technology that has impacted every aspect of society. Technology has irreversibly changed the social fabric, and it will bring far more significant societal changes in the years to come. The role of technology in shaping future societies is, therefore, multifaceted and profound, influencing various aspects of human life, economy, governance, and culture.

In the last century, technology has significantly impacted high-speed communication, high-speed transportation, biotechnology, and material science. However, their evolution was so rapid that society did not have enough time to define their path. The technologies kept evolving unprecedentedly, and the world kept adapting to them. With this trend, we may have new technologies in the future, but the existing technologies will continue to shape the world, at least for the next few decades.

Information and Communication Technology (ICT) has profoundly impacted society, transforming various aspects of human life. ICT has enabled instantaneous communication and information exchange on a global scale. People can connect across borders in real time, fostering global collaboration and cultural exchange. ICT has changed society's commercial landscape. It facilitated remote and flexible working environment-making operations at the global level, and it changed the market strategy, bringing manufacturers face-to-face with the consumers. ICT also plays an active role in shaping smart cities and revolutionizing the entertainment industry. In short, ICT has become an integral part of modern society, influencing how people communicate, learn, work, and interact with the world.

One area where ICT has played a transformative role is 'education'. ICT has revolutionized the basic approach to education. The teaching–learning model of education that existed from the inception of civilization was disrupted. The use of ICT brought education outside the formal structure of schools. It changed the education model from teaching-centric to learning-centric, from synchronous to asynchronous. Initially, there were apprehensions about this new approach to education. However, it has become far more acceptable after its usage during the COVID-19 pandemic.

While ICT offers immense potential for positive transformation, it also presents challenges and risks. Individual privacy has become a big concern, and society has become vulnerable to personal safety and antisocial activities. Societal choices, ethical considerations, and policy decisions will collectively influence the trajectory of technological development and its impact on humanity.

7 Future Universities with Current Model of Development

The future of universities is being shaped by a confluence of technological advancements, changing societal expectations, and the need for innovative approaches to education as we envision the universities of the future, several key characteristics and trends emerge, pointing towards a dynamic and transformative landscape.

In future, the one-size-fits-all model of education will likely give way to personalized and adaptive learning. Advanced analytics, artificial intelligence, and machine learning may be employed to customize educational experiences to individual student needs. Adaptive learning platforms will adjust the content, delivery pace, and assessment methods based on a student's progress and learning style, ensuring a more personalized and practical education. The rigid structures of semester-based systems may evolve into more flexible learning pathways. Modular courses, micro-credentials, and stackable certifications will enable learners to customize their educational journey. Assessment methods for a more authentic representation of students' capabilities will be evolved. Beyond traditional exams, universities would explore alternative assessment approaches, such as project-based assessments, portfolios, and real-world simulations. Competency-based assessments provide a more accurate reflection of a student's readiness for the workforce. Recognizing the importance of diversity and inclusion, future universities will strive to provide inclusive access to education. Efforts to remove barriers related to socio-economic status, geographical location, and other factors will be prioritized, ensuring that a diverse range of individuals can benefit from higher education.

The physical campus will evolve into a hybrid space that integrates seamlessly with digital platforms. Smart campuses will leverage technology for efficient resource management, enhanced security, and interactive learning environments. Physical spaces on campus will be designed to foster collaboration, creativity, and community engagement.

In the twenty-first century, technology is not just a tool but an integral part of shaping how education is delivered, accessed, and experienced. All universities or institutions of higher learning will use all or some of the following technologies.

- Digital Learning Platforms: Digital learning platforms and online education offer flexibility regarding when and where learning occurs. These platforms provide access to various courses, resources, and collaborative tools. Whether through Massive Open Online Courses (MOOCs), learning management systems (LMS), or dedicated educational apps, technology will democratize education by making it more accessible globally.
- Internet of Things (IoT): The Internet of Things involves connecting devices and objects to the Internet to collect and exchange data. IoT can enhance the learning environment in education through smart classrooms, interactive devices, and real-time equipment monitoring. IoT applications will contribute to creating more connected and efficient educational spaces.
- Virtual and Augmented Reality: Virtual and Augmented reality technologies are revolutionizing experiential learning. Immersive simulations and virtual field trips bring complex concepts to life, allowing students to explore environments that may be impractical or impossible in a traditional classroom setting. These technologies enhance engagement, deepen understanding, and provide hands-on experiences in various fields, from science and history to vocational training.
- Artificial Intelligence (AI) and Data Analytics: AI will play a crucial role in education, offering intelligent tutoring systems, automated grading, and personalized recommendations. AI-powered virtual assistants can provide instant feedback, answer queries, and guide students through their learning journey. With the help of data analytics tools, educators can gain insights into student performance, learning patterns, and engagement. By analyzing data generated through online platforms and digital assessments, educators can identify areas for improvement, track student progress, and make informed decisions about instructional strategies. This data-driven approach will enhance the effectiveness of teaching and learning.
- Blockchain for Credentials and Certifications: Blockchain technology is being deployed to secure and streamline the verification of academic credentials. This ensures the integrity of educational records, reduces fraudulent claims, and facilitates the transfer of credentials across institutions and borders. Blockchain can also support the issuance of digital badges and certificates for micro-credentials and achievements.
- Social Media and Online Communities: Social media platforms and online communities serve as valuable tools for collaboration, communication, and

knowledge-sharing in education. Educators can leverage social media for professional development, and students can engage in virtual communities to discuss coursework, share resources, and collaborate on projects. These platforms will extend learning beyond the confines of the classroom.

With the help of technology, future universities will be global hubs of collaboration, fostering partnerships with institutions, industries, and research centres worldwide. Virtual collaboration tools will facilitate real-time interactions, allowing students and researchers to work seamlessly across geographical boundaries. This global connectivity will give students diverse perspectives and opportunities for cross-cultural engagement. Innovation and entrepreneurship will be promoted on university campuses.

While universities will align with the market-driven development model, they must address sustainability concerns.

8 Future Direction for Engineering Education

Engineering education is poised for significant transformations in the coming years to meet the demands of a rapidly evolving technological landscape. Future directions for engineering education will be shaped by technological advancements, changes in industry expectations, and the need to prepare engineers for complex and interdisciplinary challenges. While the aspects elaborated for future universities will generally apply to engineering education, some special features of engineering education also need to be mentioned.

Engineers are trained to solve problems. However, social scientists pose correct problems in society. Further, real-life problems are never straight-jacketed in a particular discipline. The problems often require multidisciplinary solutions. Interdisciplinary programs and projects will prepare engineers to work collaboratively across diverse fields, addressing complex challenges that transcend traditional boundaries. The 'design your curriculum' approach may become attractive to engineering aspirants. The compulsory component of engineering courses will be reduced to a minimum with more flexibility in the choice of courses. The liberal arts component will have to be increased in the engineering curriculum to make engineers comprehend societal issues better.

The curriculum had a more hands-on approach at the beginning of engineering as a discipline. However, with rapid expansion, engineering education became more theoretical and simulation-based. In the future, the traditional lecture-based model will continue to shift towards project-based and experiential learning. Hands-on projects, internships, and industry collaborations will be essential to engineering education. This approach will reinforce theoretical knowledge and cultivate problem-solving skills, teamwork, and practical application of engineering principles. Developing entrepreneurial thinking and innovation will be critical components of future engineering education. Incubators, accelerators, and innovation hubs will be integrated into campus environments, fostering a culture of creativity and entrepreneurship. Students will have opportunities to collaborate on start-up projects, bringing their ideas to fruition. The engineers will also have to be trained to try new things, go through the less travelled path, accept failures as they come, and build on them.

As engineers must upgrade their skills for new technologies, short-term programs must evolve. Traditional degrees will continue to be necessary, but alternative forms of credentialing and recognition will gain prominence. Digital badges, certificates, and industry-recognized certifications will carry weight in the job market, reflecting the skills and competencies individuals have acquired through various educational pathways.

Overall, the future of engineering education will be marked by a shift towards interdisciplinary, experiential, and globally oriented learning. As the field continues to evolve, educational institutions, industry stakeholders, and policymakers will have to work closely to ensure that engineering education remains dynamic, relevant, and responsive to the needs of society and the ever-changing technological landscape.

9 Future Lies in Green and Sustainable Engineering

Green and sustainable engineering represents a paradigm shift in engineering, emphasizing environmental Stewardship, resource efficiency, and the integration of sustainable practices into all aspects of engineering design and processes. This approach is essential in addressing global challenges such as climate change, resource depletion, and environmental degradation. In the following, we look into some key aspects and principles of green and sustainable engineering.

Sustainable engineering adopts a life cycle perspective, considering a product's or system's environmental impacts from raw material extraction to end-of-life disposal. This holistic approach ensures that engineering decisions consider the entire life cycle, including manufacturing, transportation and usage, to minimize the overall environmental footprint. Green engineering emphasizes using renewable energy sources such as solar, wind, hydro, and geothermal. Sustainable engineering prioritizes energy efficiency and conservation as core design principles. Energy audits and modelling are commonly used tools to identify opportunities for improvement.

The choice of materials in engineering projects has a significant environmental impact. Sustainable engineering selects materials with lower environmental footprints, considering recyclability, embodied energy, and recycled content. Additionally, designing products with easy disassembly and recycling in mind supports a circular economy approach.

Sustainable engineering includes the development of smart and resilient infrastructure that can adapt to changing environmental conditions. Green building design, aiming to create energy-efficient, environmentally friendly structures and provide healthy living or working environments, is the priority for a sustainable future. This involves utilizing sensors and data analytics to monitor and optimize structural performance and incorporating climate-resilient design principles to withstand the impacts of climate change. Engineers incorporating sustainable practices recognize the importance of preserving biodiversity and protecting ecosystems.

Efficient water use and responsible water management are also critical to sustainable engineering. This includes designing water-efficient systems, implementing rainwater harvesting, and treating and reusing wastewater. Sustainable engineering also addresses water pollution and promotes conservation measures to safeguard this essential resource.

Sustainability in engineering extends beyond environmental considerations to encompass social and economic dimensions. Sustainable engineering practices prioritize community engagement, ensuring engineering projects benefit local communities and promote social equity. This involves transparent communication, understanding local needs, and incorporating social impact assessments into project planning. Sustainable engineering aligns with and often goes beyond regulatory requirements. Engineers are increasingly guided by industry standards and certifications that promote sustainable practices. Adhering to standards such as LEED (Leadership in Energy and Environmental Design) in building construction or ISO 14001 for environmental management demonstrates a commitment to sustainable principles. Many companies adopt sustainable engineering practices as part of their corporate social responsibility (CSR) initiatives. This involves setting ambitious sustainability goals (defined by the UN charter), transparent reporting on environmental performance, and engaging with stakeholders to ensure accountability and continuous improvement. Green and sustainable engineering represents a proactive and holistic approach to addressing environmental challenges while promoting social and economic well-being. By integrating sustainability into engineering practices, professionals contribute to a more resilient and regenerative future where human activities harmonize with the natural world rather than depleting it. As sustainable engineering principles become increasingly embedded in industry norms and educational curricula, the field is poised to play a pivotal role in creating a more sustainable and equitable world.

10 What Does It Take to Make Future Career-Ready Engineers?

Future-ready engineers have the skills, mindset, and adaptability to thrive in the dynamic and evolving landscape of technology, industry, and global challenges. Preparing engineers for the future requires a holistic approach beyond technical proficiency. Here are some key traits that a future-ready engineer must possess.

- Strong Technical Foundation: While non-technical skills are essential for a successful career, a solid technical foundation remains fundamental. Future-ready engineers need a deep understanding of core engineering principles, methodologies, tools, and emerging technologies. Applying theoretical knowledge to real-world problems is essential for effective problem-solving and innovation.

- Interdisciplinary Knowledge: Societal challenges often require interdisciplinary solutions. Future-ready engineers should be comfortable working at the intersection of multiple disciplines, integrating insights from computer science, biology, environmental science, and social sciences. Exposure to various disciplines fosters a holistic understanding of complex issues.
- Adaptability and Lifelong Learning: Every field rapidly increases its knowledge base, requiring engineers to be adaptable and committed to lifelong learning. Future-ready engineers embrace change, seek opportunities to acquire new skills, and stay updated on emerging technologies. This adaptability is crucial for navigating evolving industries and staying relevant despite technological disruptions.
- Critical Thinking and Problem-Solving: The ability to think critically and solve complex problems is a cornerstone of future-ready engineers. Beyond rote memorization, engineers must develop analytical skills, creativity, and the capacity to approach challenges systematically. Encouraging a mindset of continuous improvement and problem-solving enhances an engineer's ability to address novel and intricate problems.
- Communication and Collaboration: Effective communication is vital for future-ready engineers. They should be adept at conveying technical information to both technical and non-technical audiences. Collaboration skills are equally crucial as engineering projects often involve multidisciplinary teams. Engineers must work seamlessly with professionals from diverse backgrounds to achieve successful outcomes.
- Leadership and Teamwork: Future-ready engineers are proficient team members and influential leaders. Leadership skills, including motivating and inspiring others, are valuable in project management and team environments. Engineers must understand how to lead and contribute to teams with diverse skills and perspectives.
- Ethical and Social Responsibility: Engineers play a significant role in shaping the world, and future-ready engineers must recognize the ethical dimensions of their work. They should be conscious of their decisions' social and environmental impact, considering factors such as sustainability, inclusivity, and the ethical implications of emerging technologies.
- Global Perspective and Cultural Competence: Future-ready engineers should have a global perspective in an interconnected world. This involves understanding cultural nuances, appreciating diversity, and considering the global implications of engineering decisions. Exposure to international collaborations and experiences broadens an engineer's worldview and enhances cultural competence.
- Innovative Mindset: Innovation is a driving force in engineering, and future-ready engineers cultivate an innovation mindset. This involves a willingness to explore new ideas, take calculated risks, and learn from failures. Encouraging creativity and fostering an environment that values innovation contribute to developing engineers who can drive positive change.

- Digital Literacy and Technology Integration: Future-ready engineers are digitally literate and proficient in leveraging technology to enhance their work. They should be comfortable with digital tools, data analytics, and emerging technologies such as artificial intelligence, the Internet of Things, and blockchain. Integrating technology seamlessly into engineering processes is a crucial aspect of staying competitive.
- Entrepreneurial Acumen: The ability to identify opportunities, take initiative, and think entrepreneurially is a valuable trait for future-ready engineers. Whether working within established organizations or as entrepreneurs, engineers with entrepreneurial acumen can drive innovation, create value, and contribute to developing new products and services.
- Resilience and Adaptability: Future-ready engineers must be resilient in facing challenges and uncertainties. Adapting to changing circumstances, learning from setbacks, and maintaining a positive mindset is crucial. Resilient engineers persevere through difficulties and contribute to creating a culture of resilience within teams and organizations.
- Environmental and Sustainability Awareness: Engineers of the future must be acutely aware of environmental issues and incorporate sustainable practices into their work. This includes considerations for resource efficiency, waste reduction, and designing solutions that contribute to a more sustainable and regenerative future.
- Continuous Professional Development: Staying current with industry trends, advancements, and best practices is essential for future-ready engineers. Commitment to continuous professional development through workshops, certifications, conferences, and networking ensures that engineers remain at the forefront of their field.
- Community Engagement and Social Impact: Future-ready engineers actively engage with their communities, understanding the social impact of their work and contributing to solutions that benefit society. This involves considering social needs, collaborating with stakeholders, and aligning engineering projects with community priorities.

11 Redefining Universities for a Sustainable Future

The original university system evolved to develop well-rounded citizens who can create a sustainable and inclusive society. The purpose of the education system was to bring out the best in all individuals and create a happy and cohesive society. However, after industrialization, the focus of education changed to professional development. The non-professional education got a back seat. Particularly in the last century, employability became the prime focus of education. Technology started defining the educational paradigm. The materialistic growth model of development changed the societal fabric irreversibly. In the name of development, the primary value system that had survived the previous millennia was compromised. It has become abundantly clear that the current development model is neither sustainable nor desirable for humanity as a whole. Technology will do its best to address society's sustainability issues, but more than technological interventions are needed. Future citizens will have to change their thinking, and they will have to create a new (old!) value system that will make civilization perpetually sustainable.

Such a value system will transcend employability and consumerism and encompass a broader perspective on human well-being, societal progress, and the flourishing of the planet. Following are some of the dimensions of such a value system that the future university education will have to focus on in addition to the ongoing approach to education.

11.1 Holistic Human Development

Human Dignity and Well-being are the foundational aspects of a civilized society. This involves recognizing individuals' intrinsic values and dignity, irrespective of their economic productivity or consumer choices. It also entails prioritizing the overall well-being of individuals, encompassing physical, mental, and emotional health, beyond their roles as workers or consumers. Every human being has to have universal access to quality healthcare. Efforts need to be directed towards reducing health inequalities and ensuring that all individuals have the opportunity to lead healthy lives.

The health of a society does not refer to only physical health. Mental health is more important. For good mental health, community and social connections play a crucial role. Emphasis should be placed on the importance of strong communities, fostering social connections, mutual support, and a sense of belongingness. Collaboration must be favoured over competition, promoting efforts prioritizing shared goals and the common good over individual competition.

Mindfulness and present-moment awareness need to be promoted among university students. This includes promoting mindfulness practices to help individuals cultivate awareness, reduce stress, and appreciate the present moment. There is also an encouragement for a healthy relationship with technology that supports well-being rather than contributing to distraction or addiction.

Spiritual and emotional fulfilment are extremely important for creating a happy society. Diversity in spiritual and philosophical beliefs needs to be recognized and respected. Personal and societal interactions should emphasize emotional intelligence, self-awareness, and empathy.

11.2 Aligning with Nature for Sustainability

Environmental Stewardship is a core value that must be inculcated in every educated person. This includes encouraging sustainable practices that respect the environment, minimize ecological footprints, and contribute to the planet's long-term health. It also involves fostering a sense of connection to nature and understanding the interdependence between human well-being and environmental health. People will have to be trained to integrate themselves with nature. Aligning with nature automatically makes the planet sustainable.

The current development model has divided the world into two parts: the elite and the underprivileged. In an interconnected world, such a disparity makes the world unstable. Equitable distribution of resources is the key to a stable world. This includes promoting economic systems that prioritize the equitable distribution of resources and reduce disparities in wealth. Advocacy for universal access to essential needs such as food, shelter, and education should become an integral part of the education system. Support for economic models prioritizing social impact, sustainability, and ethical business practices must be provided. Work towards economic systems prioritizing shared prosperity and the well-being of all members of society will have to be encouraged. The youngsters need to think beyond the boundaries of their states and treat the world as a global family.

11.3 Personal and Professional Growth

Personal Growth and fulfilling professional experience are the keys to a happy society. For professional Growth, continuous learning is valued as an ongoing process that extends beyond formal employment, fostering a culture of lifelong learning. As emphasized earlier, the future world will be exponential with a shorter lifetime of technology. On average, professionals will have to re-skill themselves 5–6 times during their professional career. High digital literacy and up-to-date knowledge of professional requirements must be considered throughout the professional career. Self-learning habits will have to be inculcated in students of all disciplines. Universities, therefore, will have to change the teaching-centric style to learning-centric.

While there are ample training schools for professional development, hardly any institutions train in personal Growth. Individuals must be encouraged to pursue activities and passions that bring personal fulfilment and contribute to a sense of purpose. Today, an ordinary person is too occupied in professional life to give any time for his/her personal Growth. This needs to change to create a happy society. Along with their professional lives, people have to keep their brains active. A proper balance between professional and personal development will increase the happiness index of society and will reduce social conflicts. Individuals must keep some dedicated time to increase their regular interactions with family and friends. Universities will have to build a system to facilitate cohesion in the society. They must build an ethos focusing on character building to develop humans of the highest ethical standards.

11.4 Global Citizenship

Future generations must think beyond their geographical territories in the highly interconnected world. Many of the sustainability issues will have to be addressed at the global level. This involves fostering a sense of global citizenship and encouraging cooperation and mutual understanding among individuals and nations. Advocacy for collaborative efforts to address global challenges such as climate change, poverty, and inequality is to be promoted. Developing a system that encourages democratic engagement, civic participation, and social justice as fundamental components of a just and inclusive society is paramount. Support for efforts to protect and uphold human rights for all individuals should be an integral part of higher education. An emphasis must be placed on appreciating and celebrating cultural diversity and recognizing the richness it brings to society. Ethical decision-making that encourages individuals to act with integrity and empathy must be prioritized.

This expanded value system recognizes the interconnectedness of various facets of life. It seeks to foster a society prioritizing holistic well-being, social cohesion, environmental sustainability, and ethical considerations beyond employability and consumerism. It emphasizes the importance of fostering a world where individuals, communities, and the planet thrive together.

In the years to come, universities cannot remain just workforce training centres as per industry requirements. They will have to become path-defining agencies. They should not simply align with what the society or industry wants. Still, they should focus on creating thinkers who will mould the future and advise governments and policymakers on creating a cohesive and perpetually sustainable world. Universities will have to play an active role in creating a new value system. However, creating a new value system takes time and effort. It may take a few generations. The universities, therefore, will have to develop a long-term vision. A robust education in the right direction for the rest of this century may achieve our goal of realizing a perpetually sustainable, cohesive and happy world by the time we enter the twenty-second century!

Chapter 13
Reimagining Higher Education Beyond Vocational Training

Krishnasamy T. Selvan and C. Lakshmi

Abstract This chapter presents a case for reimagining the goals and approaches of higher education, primarily based on the grounds of the national purposes for higher education, models of intellectual development at the tertiary level, and the growing emphasis on student mental well-being. These overarching objectives encompass reflective thinking abilities, effective communication skills, and a positive professional and life outlook. The chapter proposes the pursuit of these goals through a well-designed curriculum comprising two primary elements: a set of courses and faculty. When taught with enthusiasm, a diverse range of subjects and knowledge systems that foster critical thinking instil similar qualities in students. While reconsidering faculty recruitment criteria is necessary, enthusiastic faculty members require committed and supportive leadership, autonomy, and encouragement of workplace spirituality. Universities are ideally positioned to provide such an environment. However, external regulatory forces, including accreditation and ranking systems which have influenced internal strategic decisions, require a comprehensive review. While considering these factors holistically may pose challenges, it presents a constructive avenue for transforming higher education and fostering the development of career-ready graduates. This type of education will also align with national aspirations and policies to cultivate a thoughtful and caring society.

1 Introduction

When education is not solely focused on preparing students for jobs but instead focuses on enabling them to seek a "higher purpose," it could accomplish so much more. The meaning of "higher purpose" has been well articulated by Drew G Faust, the 28th President of Harvard:

K. T. Selvan (✉)
Department of Electronics and Communication Engineering, Sri Sivasubramaniya Nadar College of Engineering, Kalavakkam, India
e-mail: selvankt@ssn.edu.in

C. Lakshmi
Department of English, College of Engineering Guindy, Anna University, Chennai, India

I. Gratchev et al. (eds.), *Transformative Approaches to Career-Ready Education*,
https://doi.org/10.1007/978-981-96-7018-5_13

> Higher learning can offer individuals and societies a depth and breadth of vision absent from the inevitably myopic concept. Human beings need meaning, understanding, and perspective, as well as jobs. The question should not be whether we can afford to believe in such purposes in these times but whether we can afford not to. (Faust, 2009)

There are connections between career preparation and the higher goals of education. While a job involves useful work and provides a livelihood, the word *career* represents something more significant. A career can represent a "path or progress through life or history" and a "profession or occupation chosen as one's life's work." A career is a *life-long process* (Collins English Dictionary). Thus, career readiness means being prepared for professional pursuit *throughout* one's life.

Job-readiness, often strongly emphasized (Moore & Morton, 2015), refers to the case where education is implemented essentially to help graduates secure a job immediately after the completion of university. When job preparation is the main focus, it can be expected that there are two primary goals for educators:

(1) To strive to improve learners' subject proficiency
(2) To ensure skill learning that helps learners contribute to their immediate jobs

Establishing a learning foundation through subject knowledge and related skills is an essential teaching goal. However, exclusively acquiring subject knowledge does not prepare the learners for the challenges of a fast-changing social, cultural or technological milieu. If education is "tailored too closely to particular jobs or workplaces, it becomes inflexible," as the skills are not readily moved to other places (Moore & Morton, 2015). Rapid industrialization and 'internetization' (Vass, 2017) over the last couple of decades have made us aware that graduates need many additional skills for a fulfilling career. Since the type of the specific skills often changes from industry to industry and from time to time, the modern university is tasked with providing a generic set of skills along with a strong foundation in the core subjects that will help the graduates *adapt* to varying industry requirements.

Based on a consideration of six perspectives, this chapter argues for broadening job-readiness goals to include personal growth and enrichment of the students, nurturing enthusiastic young individuals who are *reflective thinkers* and effective communicators capable of facing setbacks or failures without breaking down. Reflective thinking, central to many other qualities, involves carefully weighing options and reasoning before making decisions or forming beliefs. A comprehensive prescriptive model for reflective thinking encompasses problem identification, considering alternatives, logical reasoning, adaptation, and critical evaluation. Reflective thinking allows the learner to "choose courses of action" that lead to achieving goals that may be otherwise unattainable (Baron, 1981).

There is a need for a realigned worldview that includes the teaching of reflective thinking, openness, lifelong learning, and the development of a positive outlook as a comprehensive part of higher education (HE) (Bauer, 1992; Candy, 1995). Interestingly, once thus expanded, the broader set of goals becomes consistent with the higher purpose and the career-readiness purpose of higher education.

The educator today has the challenge of integrating "innate" goals with traditional goals, which, the authors believe, demand more enthusiasm and commitment. The environment that the teacher creates in the classroom is also essential to making the goal-broadening exercise a success. By holistically broadening educational goals, higher education becomes the stepping stone that advocates students to be more proactive and engaged in lifelong learning.

Of course, such a higher education is only possible if the ecosystem is sufficiently supportive. Therefore, this chapter addresses the necessary aspects of several essential components of the ecosystem—leadership, university culture, and regulatory and ranking systems.

While we aim to provide specific ideas for implementation, we will also share thoughts in some cases to contribute to continuing dialogues on this important topic of higher education, which will always be a work in progress.

While there are varied interpretations of the terms 'university' and 'higher education' (Barnett, 2004), in the context of our discussions in this chapter, we refer to physical campuses that primarily offer in-person education, regardless of whether research is a significant part of their activities.

2 Grounds for Reimagining Educational Goals

We now present six grounds, some of which might have overlapping considerations, which compel us to consider expanding our educational goals centred around job readiness.

2.1 What Are Universities for?

Scholars have long reflected on the purpose of university education.

If higher education exists to acquire new knowledge and establish a great workforce, could it not simultaneously strive for higher ideals and goals? Chan (2016) suggests two purposes that go beyond the "economic instrumentality" of higher education:

- Creating an engagement that is democratic, civic-centred and addresses real-world problems
- Developing a fully rounded, intellectually developed and caring person (Chan, 2016)

Sarvepalli Radhakrishnan, a great philosopher and educator and the second President of independent India, stated that the aim of education includes the training for democracy and self-development and the development of qualities of mind like fearlessness, the strength of conscience, integrity and lifelong learning (Debnath, 2019). Boulton and Lucas (2008) have argued that the university is "the only place where that totality of ourselves and our world is brought together and which makes it the strongest provider of the rational explanation and meaning that societies need."

They go on to say that such “statements about the deeper values of education” are not merely “sentimental” but are “deeply utilitarian.”

2.2 National Aspirations for Higher Education

Higher education reflects a nation’s aspirations for its young minds brimming with potential and promise. It is not merely a tool for achieving material success but a provider of hope for a better future for all living beings. Each nation sets its educational goals and purposes for higher education. In this section, we review some of these goals documented in official government policy papers from India, Australia, and the U.K.

In the case of India, the 2020 National Education Policy (NEP, 2020) states that the purpose of higher education should centre around three aspects: (1) Employment, (2) Personal Development and (3) Positive outlook. More specifically, higher education must be designed to

- contribute towards sustainable livelihoods and economic development
- develop good, thoughtful, well-rounded and creative individuals
- prepare students for more satisfying work and personal lives
- create a happier and more cultured society

In order to work towards the above, NEP advocates the need to adopt a more holistic educational perspective by

- Moving less towards content and more towards learning how to learn
- Creating a more liberal educational structure inspired by ancient Indian universities[1]
- Facilitating the development of intellectual, aesthetic, social, physical, emotional and ethical capacities of learners

A white paper by the National Accreditation and Assessment Council of India (Patwardhan & Mohanan, 2022) also highlights the need to go beyond information transfer and skills acquisition to *education of the mind*. The paper argues that universities should work towards *educatedness*, which is defined as a set of intelligences that empower individuals to strive for their well-being and that of others (Patwardhan & Mohanan, 2022). To this end, the paper recommends, in consistency with NEP, that higher education should comprise two components:

(1) A general education focusing on cognitive development.
(2) A specialized education focused on the areas of the chosen discipline of the learner.

[1] At the ancient Nalanda University in India (600 A.D.–700 A.D.), students from different parts of India, China, and Malaysia learned Grammar, Science, Engineering, Medicine, and Philosophy (Chambers 2017; Moorkerji, 1941; Pinkney, 2014). Thus, this ancient Indian university adopted a form of liberal education.

A government-constituted committee published the "Higher Education in the Learning Society" report in the United Kingdom in 1997. According to the report, the purpose of higher education includes:

- Developing capabilities of individuals to the highest potential
- Knowledge and understanding for their own sake
- Enabling effective contributions to society and personal fulfilment (Dearing, 1997)

A higher education report from Australia suggests contributing to social and economic change within the nation, in addition to producing and disseminating new knowledge and applying that knowledge, as the main functions of higher education. Higher education must also:

- Promote a deeper understanding of health and social issues, and reduce social and economic disparities.
- Foster international engagement, creating a confident and globally aware nation.
- Sustain and rejuvenate other societal institutions.
- Fulfil a traditional role in personal development, transforming individuals' lives and fostering a love for learning and intellectual discovery (Review of Australian Higher Education, 2008)

These national aspirations towards higher education are humanistic and comprehensive. They expect universities to prepare students for achieving broader objectives aligned with a higher purpose. As can also be inferred, the national aspirations are consistent with the purpose of higher education as articulated by scholars (Sect. 2.1).

2.3 Theories and Models Exploring Learner Development

Several theories have been proposed with respect to learner development. They analyze how learning occurs at the university level and the processes the learners experience to achieve self-efficacy and learner autonomy. This section reviews the important aspects of three models of intellectual development (Ebel, 2009):

(1) Perry's Scheme of Intellectual Development
(2) Baxter-Magolda's Model and
(3) the King-Kitchener Model

Perry's intellectual development model, developed in the 1970s, explains how individuals' views shift and evolve, based on studies conducted with undergraduate college students. Perry describes four stages in this journey: dualism, multiplicity, relativism and commitment (2003). Learners begin their journey with the worldview that every problem has a right solution. However, they eventually understand that multiple answers may exist, but not only that, there could also be unsolvable problems. Personal perceptions will matter in finding those solutions. Relativism is the stage in their intellectual development where learners understand the significance of contextual knowledge. Finally, in the fourth stage, learners understand how to integrate knowledge learned from others and combine it with their learnings and experiences, thus committing to an understanding of ambiguity and uncertainty that is part of the process of learning, as opposed to the start of their journey where duality dictated understanding (Cassidy, 2004; Ebel, 2009).

Baxter-Magolda's Epistemological Reflection Model of intellectual development is especially suitable for understanding different types of learners. *Absolute knowers* view knowledge as certain; transitional knowers assume knowledge is absolute in some areas and uncertain in certain others; *independent knowers* understand that knowledge is mostly uncertain, and *contextual knowers* believe knowledge is necessarily uncertain, but context can decide which knowledge is better than the others (Hettich, 1998; Ebel, 2009). At this level, learners exhibit openness, lifelong learning, critical thinking and reflective thinking.

The King-Kitchener Model, which has seven stages, identifies reflective thinking as the highest form of intellectual development, as learners understand how to process all evidence available and arrive at judgements with better efficiency (King & Kitchener, 2004).

Evidently, all of these intellectual development models have a similar understanding of reflective thinking as a higher form of intellectual competency and learner autonomy.

All of the models show why a singular attention on development in subject knowledge and narrowly related areas does not necessarily produce the highest level of intellectual development for anyone who partakes in higher education. The models are similar in their conclusion that learning is a lifelong, reflective journey that ought to change a person for the better. Learners start with many assumptions about their subjects, college lives, and their selves and the outside world. These assumptions get challenged, refined, corrected or cemented as they advance from one semester to the next through interactions with their faculty, peers and society at large. Higher education institutions and teachers are responsible for creating an environment where this capacity for reflective thinking can develop continuously.

2.4 Students' Views on the Purpose of Higher Education

As major stakeholders, what do students expect from their higher education experience? It may be a reductive assumption that students are concerned about employability and do not give importance to the higher purposes of education. We reviewed research papers that surveyed student perspectives on higher education which presented illuminating results.

Brooks et al. (2020), analyzed students' perspectives from six European countries to understand what they considered the purpose of higher education. According to their study, students display a nuanced understanding of the purpose of higher education. They express the opinion that higher education must not be focused solely on the job market but also on 'personal growth and enrichment, as well as societal development and progress.'

Higher education also ought to facilitate intellectual inquiry, which is of intrinsic worth. Students believe that higher education should provide "critical engagement" that "promotes collective good" to make young people "curious and inspired." "It is not necessarily about getting great grades but about learning from your mistakes and becoming more resilient" (O'Shea & Delahunty, 2018).

In a soon-to-be-reported international study conducted in five different institutions in Australia, Brazil, Canada and India among a total of over 900 undergraduate students, overall, the respondents chose the development of personality and a positive approach to profession and life as the second and third most important purposes of education, closely following employability (Selvan et al., 2025).

Thus, students want to be inspired and motivated, in addition to being trained in skills and techniques. We need to consider whether the current goals and approaches that most higher education institutions work hard to accomplish, invoke creativity and imagination of twenty-first century learners.

2.5 *Addressing Challenges in the Modern World*

Resolving contemporary challenges requires reflective learning ability. Reflective learning is closely related to creative thinking (Akpur, 2020) and, hence, innovation and research capacity building. Moreover, it also facilitates the related qualities of professional development, openness, and lifelong learning and aids in developing competencies that may help promote sustainability (Colomer et al., 2020).

Technological advancement has meant easy access to "rapidly disseminating information" (Akpur, 2020). Therefore, education must help students make "clear, reasoned judgements based on interpreting, understanding, applying and synthesizing" (Burns & Sinfield, 2016) from the nearly infinite amount of information available on any given topic. Hence, students must learn to discern what kind of information to consider for further examination and what to discard (Osborne & Pimentel, 2022). To do so, they must acquire the intellectual tools to identify and dismantle misinformation. Thus, university education is confronted with the following questions:

- How should students make sense of too much information on the internet or anywhere?
- Can educators help students distinguish between the true and the seemingly true?

It is worth recollecting a most appropriate couplet from Thirukkural, a classic Tamil language text written by Thiruvalluvar reportedly between 500 BCE and 5CE (thirukural.net, 2021),

> the mark of wisdom is to discern the truth
> from whatever source it is heard. Venkatachalam (2015).

2.6 The Critical Issue of Student Mental Well-Being

Mental well-being is integral to a student's academic growth and overall development, essential for becoming a well-adjusted member of their chosen profession and society. Higher education policies in many institutions insist on the presence of counsellors, mental health advisors, and other similar initiatives to achieve this. However, these measures are not overarching and have yet to make a serious impact globally.

The relationship between learning and mental health has already been established in the literature, with several studies showing a direct link between motivation, self-efficacy, learner autonomy and the mental well-being of the students (Houghton & Anderson, 2017). Focusing on student well-being measures can also bring many other benefits and rewards to the institution's stakeholders. Research has proven that an improved mental state, in turn, improves academic learning and development (Hammond et al., 2019).

The United Nations ranks suicides as the third leading cause of death among the young population (WHO, 2023). Many student suicides in India are reported by the top institutions (PTI, 2023). The U.S. reported the highest number of suicides since 1941 in 2021 (Shrikant, 2023). Shrikant reports that the increase is observed both in the country's affluent and non-affluent student communities.

The issue of student mental well-being must be widely discussed, and efforts must be made to make it an integral part of the higher education curriculum. Of course, there are barriers (Houghton & Anderson, 2017):

(1) A lack of widespread discussion on mental health and well-being in connection with the teaching–learning context
(2) Assumption that the college campus is providing sufficient support in appropriate ways to all the students that need it
(3) Mental health challenges that academics face in high-pressure environments

Mahler et al. investigated the relationship between teacher enthusiasm and student performance and found a positive association between teachers' subject-specific enthusiasm and student performance (Mahler et al., 2018). Hence, when teachers exhibit exuberance and enthusiasm and model themselves as life-long learners, it can positively impact student learning and their mental well-being. In fact, we often observe students develop professional identities aligned with the values teachers demonstrate (Selvan, 2023). Therefore, as much as imparting subject knowledge, it may help for the teachers to demonstrate a positive, enthusiastic approach to improving students' mental health.

2.7 *Summarizing Grounds for Broadening Educational Goals*

By synthesizing the discussions in the above sections, we can summarize the grounds for broadening goals for university education:

- Graduates may be placed in different organizations, not necessarily in the fields or the specializations in which we prepare them. As stated by Moore and Morton, 2015, "it is not at all possible to generalise about the specific types of workplaces and workplace roles that our graduates will ultimately enter." Hence, a narrow focus on gaining subject knowledge may be less rewarding than a broader set of goals for higher education.
- Discipline-specific skills in many areas have only a short life, and what will be needed even in the medium term cannot be predicted with any precision.
- The quality of learners' intellectual development depends on critical thinking ability, which is crucial to fostering creativity and innovation.
- Students desire their universities to have a broader purpose.
- Students' mental well-being has far-reaching personal and societal implications.
- We must relentlessly pursue the articulated national purpose of higher education in letter and spirit.

The challenges of the twenty-first century are too many, and higher education systems the world over have perhaps been caught underprepared or underequipped to meet them. We propose that education goals must be reset.

3 Goals for Higher Education Beyond Job Readiness

Apart from subject knowledge and skill development as the primary goals, we need to further broaden the goals to include the development of:

- Reflective thinking ability, openness and lifelong learning
- Creative thinking and research skills
- Effective language and communication skills
- Personality and a positive approach to profession and life

The importance of these goals is evident: Our graduates have to be reflective thinkers to make self-assessments about their learning and understanding of the world. Reflective thinking can also help them be innovative and creative in addressing or facing modern technological and societal challenges. Similarly, language and communication skills are required to make effective presentations, write well, and even correctly understand lectures and textbooks; they are also fundamental skills that allow us to make human connections. As educators, these are the qualities we would like to see in our students. Should we not want our students to be positive, enthusiastic and cheerful while engaging in their professions? Should we not help them develop the ability to manage setbacks or failures and face challenges in life without undue fear? How can educators and universities and the ecosystems they are part of facilitate all these qualities?

4 How Do We Integrate the Broader Goals in University Education?

The curriculum, as the primary means for the university to integrate broader goals, is a comprehensive entity. It's essential to understand that the academic curriculum in higher education is a culmination of lived learning experiences from the syllabi, interactions with the faculty, the classroom and college campus environment, and the culture that is built within. While the curriculum is a complex interplay of several components, two components, namely the set of courses and the teacher's persona, stand out in terms of their importance. However, it's worth noting that the support systems, facilities, and educational resources also play a crucial role in the curriculum, albeit in a supporting capacity.

4.1 Liberal Education Model to Shape Curriculum Design

As we have stated already, the development of the capacity to reflect and learn how to learn are foundational objectives of higher education. In India, NEP 2020 recommends bringing back the model of liberal education to facilitate these capacities (NEP, 2020).

What is liberal education? According to Mulcahy,

> ...the ideal of the educated person has come to mean a person of intellectual formation, one who possesses knowledge in depth and breadth, one who possesses the knowledge and skills of citizenship, and who is respectful to others and caring towards them, and one who is enabled to engage in thoughtful action (Mulcahy, 2009).

To facilitate such a development, liberal education must include an extended range of subjects and knowledge systems that help develop skills such as creative thinking, teamwork, problem-solving, civic knowledge and engagement, ethical reasoning and action, and synthesis and advanced accomplishment across general and specialized studies (Mulcahy, 2009; Strauss, 2011; Scott, 2015).

Fox (2016), lists the ability to speak and write clearly and think critically as integral components of a liberal education, arguing that rather than mastering specific content knowledge, liberal education must attempt to provide an approach to learning. General education is one such approach in which "learning methods are emphasized over specific content" (Fox, 2016). Such an education can help with cognitive development and creative thinking ability. Students must also gain statistical literacy, information literacy, moral and civic education, and skills for living in a globally diverse world and lifelong learning through a liberal education.

The core values sought to be instilled by general education should also be carried through to specialized education so that students do not experience a disconnect (Bourke et al., 2009).

If the set of courses represents the body, the teacher is its soul. Only with effective and enthusiastic delivery, appropriate assessments, and other related aspects can any set of courses have a meaningful positive impact. This brings the teacher's motivation and enthusiasm into focus.

4.2 The Centrality of the Teacher Motivation and Enthusiasm

Learner outcomes are strongly impacted by teacher enthusiasm. As Sarvepalli Radhakrishnan said, university teachers must embody the highest level of intellect, enthusiasm and cheeriness (Debnath, 2019). Based on a review of the literature, research shows that an enthusiastic teacher makes valuable contributions towards student learning. When teachers are enthusiastic, learners show more interest, on-task behaviour, motivation and recall (Buric & Moe, 2020).

One of the most observable characteristics that effective and successful teachers have is a passion for the subject. Teachers' enthusiasm towards the subject can translate to learner enthusiasm in the classroom. Teacher enthusiasm can be of two types: "Experienced enthusiasm", which includes the enjoyment and excitement teachers experience while at work, and "Displayed enthusiasm" which is visible to others through their positive body language and expressive behaviour (Keller et al., 2015). Teacher enthusiasm invigorates student interest and excitement (Patrick et al., 2000), and in turn, enthusiastic and motivated students inspire teacher enthusiasm.

Enthusiastic faculty are pivotal in facilitating a rich educational experience for students. They have direct and indirect impacts on students, encouraging them to develop a positive approach to their profession and life. This, in turn, complements a well-designed syllabus, engaging young learners in a holistic learning process. In the soon-to-be-reported study noted in Sect. 2.4, the students ranked 'vibrant teaching and learning' overall as the primary facilitator for achieving educational goals, highlighting their active role in the learning process.

One of the main reasons for consistent, persevering faculty enthusiasm is intrinsic motivation—interest, satisfaction, and fulfilment in what one does that is not reliant on rewards or external validation. Extrinsic factors also impact faculty enthusiasm which should be considered so that universities can create an atmosphere where teachers and learners thrive.

5 Extrinsic Factors Affecting Faculty Motivation

While intrinsic motivation factors are valuable, positive emotional experiences in the work environment create enthusiastic teachers, increasing teacher efficacy and job satisfaction (Buric & Moe, 2020). According to Russell, "…a good alignment between an individual's job responsibilities, his or her mission, and the mission of the organization or work unit can lead to passion in work" (Russell, 2008).

Selvan (2023) identifies the following extrinsic factors as being important:

- Salary, autonomy, and supportive workplaces
- Elevating the status given to teachers by recognizing their role in letter and spirit in shaping future citizens

Aligning these extrinsic motivators with intrinsic factors can cultivate passionate teachers, benefiting both the profession and society.

It is refreshing that NEP (2020) recognizes the importance of placing the faculty at the centre of educational reforms. Given that low faculty motivation is a significant challenge in higher education around the world, this recognition is essential. In the Indian tradition, the teacher or 'the guru' is esteemed and valued. Today, there is a need to revitalize the teaching community as respected and essential members of society who are happy, enthusiastic and motivated (NEP, 2020).

Daumiller et al. (2020), remark on the 'underdeveloped' nature of research on faculty motivation and the importance of recognizing faculty emotions as an integral component of teaching. According to them, while faculty performance is critical to society, teaching almost consistently gets de-emphasized over research expectations. Job satisfaction and faculty well-being are also relegated to low-priority items. "Faculty emotions have [also] been identified as unique across work domains, with research typically yielding more negative and teaching more positive emotions, which also correlate with perceived success". Teachers' emotions, their sense of accomplishment, and the ability to regulate their emotions can influence the classroom environment as well as the students' emotional experiences.

Extrinsic motivating factors are primarily shaped by the immediate ecosystem that comprises the internal systems of leadership and organizational culture. These systems are, in turn, affected by the external forces of regulatory and ranking systems. The collective influence on faculty, teaching and learning (read students), and research quality is rather significant. Therefore, it is crucial to emphasize the need for a comprehensive consideration to contemplate the substantial positive changes that are needed.

6 The University as a Nurturing Workplace

6.1 Why Leadership Matters

As Nathan (2022) noted, a sense of purpose truly motivates people and helps them be committed. Commitment "really comes from within." Therefore, a leader needs to make efforts to develop a sense of purpose amongst the organization's employees so that this commitment becomes a culture of the organization.

A university's standing is a reflection of its teaching and research quality. These, in turn, are primarily administered by the body of its faculty. The implication is that every teacher represents, or ought to represent the university and its vision, and therefore is a leader on their own. The university leadership has to facilitate an environment and work culture that promotes this identification with the university's vision. This necessitates motivated and visionary leadership. It is, therefore, not surprising that NEP talks of the need for outstanding and enthusiastic institutional leaders to cultivate excellence and innovation in higher education institutions (NEP, 2020).

Promoting enthusiasm at work enhances productivity and well-being. It requires a collective effort, with all members reflecting on motivations and prioritizing others' needs alongside tasks. Educational leaders play an essential role in fostering environments where enthusiasm and cooperation can thrive. Achieving a balance between collaborative leadership, strong work ethic and mission alignment is challenging but worthwhile (Russell, 2008).

Higher education needs dynamic, transformative and collaborative leaders—those who can connect to the organisation's members and successfully build consensus (Delener, 2013). Higher education leaders must be capable of balancing tradition with change, necessitating effective communication and authentic stakeholder relationships (Delener, 2013; Lambert, 1998). Overall, successful leadership in higher education requires vision-setting, community-building, empowerment, and strategic implementation alongside a broad knowledge base encompassing pedagogy, innovation, global influences, leadership capacity-building and crisis communication (Chenoweth & Everhart, 2002).

6.2 Promoting Trust as a Value

An educational institution functions on trust—there is, or there ought to be, trust between students and teachers, between teachers and administrators, and between administrators and students. Thus, by extension, there ought to be trust between the university and all its stakeholders. This trust is essential to the institution's functioning and, in fact, to its very existence (Zalec, 2014).

Universities are intensely social organizations and are therefore accountable to society and its various stakeholders. Accountability is the act of demonstrating trustworthiness. While at some level this is required, in recent years, there have been increased auditing practices in institutions, implying an excess of inspections, documentation and filing of evidence. These actions have directly and indirectly resulted in controlling measures such as over-standardization, top-down auditing approach and reduced reliance on self-evaluation, all of which Cowie and Croxford label as non-intelligent accountability in education (Cowie & Croxford, 2007). One of the possible side effects of this "non-intelligent accountability" is its negative impact on faculty.

The one main reason that can account for the increased auditing requirements is the lack of trust in our systems and peoples. There is an urgent need to improve trust levels in our society, and universities are the ideal starting point. Therefore, every institution should promote the value of trust in suitable ways. Without trust, excellence may not follow.

6.3 Workplace Spirituality as a Positive Force at the University

As discussed, we expect our students to develop as enthusiastic individuals with a positive outlook, which calls for faculty enthusiasm. The combination of enthusiasm and positive outlook lends well to the term "workplace spirituality"—a management concept that can aid university leadership in making the campus a reasonably pleasant workplace.

Workplace spirituality is not about a religion or a particular socio-cultural practice. Fernandes Bella et al. (2018), describe it as a "human-centred view based on self-awareness". It is "an individual's attempt to live his/her values more fully in the workplace." The concept has three dimensions:

- Inner life or spiritual identity—What do I want to do?
- Sense of purpose—What do I align with?
- Sense of community or connection—Whom do I share common interests and ideas with? Do my values align with those of my organization?

Hence, fostering meaningful connections in the workplace is crucial to gaining a sense of professional satisfaction and achievement. Enjoying one's work, establishing a sense of connection with co-workers, and feeling aligned with the objectives and philosophy of the organization are important to realizing this. A study examining workplace spirituality and subjective happiness among high school teachers found that these factors have a significant relationship with one another, with gratitude acting as the moderator (Mahipalan, 2019). Another study conducted in Indonesian universities involving 322 lecturers found a positive relationship between workspace spirituality and employee engagement (Margaretha et al., 2021). Therefore, for teachers who help shape the future generation, and the leadership, workplace spirituality in the university is desirable and we must nurture it.

7 The External Factors

Many of the policies and approaches of the university leadership are often shaped by pressures from the external regulatory and ranking systems. We briefly review these factors and share desired approaches moving forward.

7.1 Regulatory Systems to Promote Excellence

Regulatory systems can be heavy-handed and do not necessarily encourage autonomy (NEP, 2020; Selvan and Warnick, 2023).

Selvan and Warnick reviewed the process of engineering programme accreditation and stated that the current accreditation procedures rely heavily on documentation, risking administrative overload and impeding progress. Despite "neither the learning process nor its outcome in the mind" being "directly observable" (Patwardhan & Mohanan, 2022), much of these efforts are directed towards measuring learning outcomes, which does not appear to represent time well-spent. Moreover, it makes it possible for programs to resort to superficial adjustments or excessive paperwork to show on paper that standards are met, neglecting genuine enhancement (Selvan & Warnick, 2023). While we must expend efforts towards desired learning outcomes, any efforts to measure the *holistic* development of students, a practice that has been in vogue for several years, is fraught with the risk that the results may be misleading (Williamson et.al, 2020). For example, while an Australian-based study noted that graduates have been found lacking in the highly required skill of written communication (Moore & Morton, 2015), NEP flags "less emphasis on the development of cognitive skills and learning outcomes" as a primary concern in higher education in India (NEP, 2020). The reduced emphasis on learning outcomes can also affect research quality.

An effective accreditation framework should be based on trust and prioritize educational excellence. It should offer flexibility in course topics and qualitative assessments. Additionally, it should empower faculty and promote innovation. Thus, accreditation must evolve to emphasize quality, innovation, and faculty autonomy, ensuring an enriching educational journey for students (Selvan & Warnick, 2023).

We need a balanced approach that values both standardization and flexibility, allowing institutions to uphold their unique educational missions while meeting essential quality benchmarks. Ultimately, accreditation should serve as a real catalyst for continuous improvement.

7.2 Deprioritizing Ranking Systems

Let us consider the student assessment of teaching based on numerical scores. For example, what does this convey to us if we see a score of 7 out of 10? There is nothing much except that the teaching is assessed to be reasonably good. We do not know which teaching component contributes to its "being good" in this case. On the other hand, a qualitative statement such as "the teaching was inspirational" conveys much richer information. In general, qualitative methods can "provide rich descriptions of complex phenomena." At the same time, dependence on only quantity leads to a risk of "ignoring factors that are more significant in explaining important realities and relationships" (Sofaer, 1999). Universities are, undoubtedly, highly complex organizations and are not amenable to simple quantification by any means. The only way it can be done, as appears to be the case in the current ranking systems, is by ignoring several "higher education values to be exalted," such as the training of "good professionals" (Vidal & Ferreira, 2020).

National and international university rankings are generally "derived from a set of underlying quantitative data" (Johnes, 2018) and, therefore, are of questionable "validity and reliability." Naturally, these rankings have "weaknesses," besides posing some "threats." The threats include (Vidal & Ferreira, 2020):

- Ignoring "many important aspects related to the quality of university education…such as human capital formation," and the conversion of the "means (measurement) into the goal (to be measured)"
- A doubt about "what is really being measured"
- Annual publication of thc rankings

The third threat creates pressure on institutions—their leadership and faculty—leading to "wasting time and money explaining their rank in every ranking published yearly" rather than focusing on their mission (Vidal &Ferreira, 2020). Universities must decide whether to conform to rankings standards or maintain their unique identities despite media pressures. Chasing rankings may have to be deprioritized if universities were to focus on their missions and transparently communicate accomplishments to stakeholders (Vidal & Ferreira, 2020).

8 Discussions and Key Recommendations

Job readiness is a short-term goal, so other factors are needed when defining educational objectives for universities. Career readiness and the higher purposes of education focus on the long-term development of graduates and mutually reinforce each other. Efforts towards reimagining goals and approaches with either of these purposes as the primary goal may lead to more satisfactory individual and collective development.

Any attempt at a truly collective development must acknowledge the presence of socio-economic, academic, and institutional barriers that preclude marginalized or underrepresented groups from higher education. Financial support in the form of scholarships alone cannot ameliorate the underrepresentation of these communities in our universities. Students who belong to diverse socio-cultural and economic backgrounds can thrive in a liberal education environment that provides strong mentorships and a network of support systems. A holistic higher education philosophy that is not laser-focused on standardized clinical examination results to measure students' abilities will also promote better integration and wider participation of students from all sectors of the society.

While plans to facilitate the holistic development of students are important, it is the faculty that gives *life* to the contents. So we must examine ways of promoting faculty enthusiasm and motivation. Since university culture and leadership impact faculty motivation and educational outcomes, these need to be considered along with the external forces that pressure university leadership. In the interests of higher education and society overall, we should examine all these aspects holistically and not in silos. National and international professional societies, as well as universities themselves, can steer dialogues on these aspects. A common refrain may be that such an effort will be too challenging. However, if students are expected to tackle challenging problems, can the collective body of professionals, universities and governments not tackle complex problems as well? Our students and society deserve this effort.

Some key recommendations that follow from the discussions in this chapter are:

(a) Curriculum design should be shaped on the model of liberal education.
(b) Constant efforts are required to make universities great places to work so that they can attract and retain the best minds. To this end, besides taking care of the obvious factors such as autonomy and a respectable salary, institutions should promote trust and workplace spirituality on their campuses.
(c) Inclusion of the following eligibility criteria in addition to the existing norms in faculty recruitment should be considered: a commitment to the purpose of education and the institution's values, enthusiasm and an interest in contributing to the development of students. Given the dynamics of knowledge and technology systems, mere scholarship in a particular field of research may not help a young faculty remain motivated throughout a long academic career.
(d) Establishing faculty development units in universities that address aspects of teaching, assessment, research, professional development and motivation may help nurture a continuously reflective ambience.
(e) National and international professional societies may run leadership development programs to mentor academic leaders. These programs cannot adopt a one-size-fits-all approach but instead be tailored to the needs of specific university types, as different types of institutions face different challenges (with some common ones, of course). Accreditation requirements may warrant senior leadership attending at least one such program yearly.
(f) Accreditation and ranking must be the byproducts and not the end-goals of the work done by an institution towards fulfilling its mission. They should

not require undue activity from faculty and staff. With similar reasoning, while workshops on teaching and learning, research, professional development and motivation are welcome, over-reliance on workshops for accreditation or rankings may suggest flaws in the regulatory and ranking systems.

(g) We should encourage a culture where universities may choose to ignore ranking systems and instead focus on their mission and providing "valid and reliable information to all stakeholders about the level of achievement of their goals" (Vidal & Ferreira, 2020)

Academics, university leaders, professional societies, and governments can profitably reflect and engage in dialogues on these critical factors affecting higher education and our collective future. In this process, it is crucial to recognize that "too much authority and expertise" and decision-making are not left "to the predilections and interests" of "external, non-educational agencies" (Moore & Morton, 2015). Universities, as the key players, must remain, or become, places where teaching and learning, research and professional development activities are joyfully pursued to maximise their efficiency and social contributions, and hence our collective betterment.

Acknowledgements The authors most sincerely thank Karl F. Warnick and S. Shyam Krishna for their edits and suggestions that helped make this chapter better.

References

Akpur, U. (2020). Critical, reflective, creative thinking and their reflections on academic achievement. *Thinking Skills and Creativity*, 100683.

Australian Government. (2023). *Australian Universities Accord: Summary of the Final Report.* Department of Education, Australian Government.

Barnett, R. (2004). The purposes of higher education and the changing face of academia. *London Review of Education*, 61–73. https://doi.org/10.1080/1474846042000177483

Baron, J. (1981). Reflective Thinking as a Goal of Education. *Intelligence*, 291–309.

Bauer, N. J. (1992). Dewey and Schon: An analysis of reflective thinking. In *Annual Conference American Education Studies Association* (pp. 1–31). American Educational Studies Association.

Bloomfield, G. (2016, April). Why do engineering graduates choose non-engineering careers? *Engineering and Technology Jobs.*

Boulton, G., & Lucas, C. (2008). *What are universities for?* Retrieved from https://www.leru.org/files/What-are-Universities-for-Full-paper.pdf

Bourke, B., Bray, N. J., & Horton, C. C. (2009). Approaches to the core curriculum an exploratory analysis of top liberal arts and doctoral-granting institutions. *The Journal of General Education*, 219–240.

Bradley, D. (2008). *Review of Australian Higher Education.* Commonwealth of Australia .

Brooks, R., Gupta, A., Jayadeva, S., & Abrahams, J. (2020). Students' views about the purpose of higher education: A comparative analysis of six European countries. *Higher Education Research & Development*. https://doi.org/10.1080/07294360.2020.1830039

Buric, I., & Moe, A. (2020). What makes teachers enthusiastic: The interplay of positive affect, self-efficacy and job satisfaction. *Teaching and Teacher Education*, 1–10.

Burns, T., & Sinfield, S. (2016). *Essential study skills: The complete guide to success at university.* Sage.

Candy, P. C. (1995). Developing lifelong learners through undergraduate education. In L. Summers (Ed.), *A focus on learning: Proceedings of the 4th annual teaching learning forum* (pp. 2–7). Edith Cowan University.

Cassidy, S. (2004). Learning styles: An overview of theories, models,and measures. *Education Psychology*, 419–414.

Chambers, C. R. (2017). Discovering Nalanda and other institutions of higher education before Solerno. *Journal of the Study of Postsecondary and Tertiary Education*, 196–205.

Chan, R. Y. (2016). Understanding the purpose of higher education: An analysis of the economic and social benefits for completing a college degree. *JEPPA*, 1–41.

Chenoweth, T. G., & Everhart, R. B. (2002). *Navigating comprehensive school change: A guide for the perplexed.* Eye on Education.

Collins English Dictionary. (n.d.). Retrieved March 29, 2024, from Collins: https://www.collinsdictionary.com/dictionary/english/career

Connel, R. (2019). *The good university.* Bloomsbury Academic.

Cowie, M., & Croxford, D. T. (2007). Tough, intelligent accountability in Scottish Secondary Schools and the role of Standard Tables and Charts (STACS): A Critical Appraisal. *Scottish Educational Review*, 29–50.

Colomer, J., Serra, T., Cañabate, D., & Bubnys, R. (2020). Reflective learning in higher education: Active methodologies for transformative practices. *Sustainability, 12*(9), 3827. https://doi.org/10.3390/su12093827

Daumiller, M., Stupnisky, R., & Janke, S. (2020). Motivation of higher education faculty: Theoretical approaches, empirical evidence, and future directions. *International Journal of Educational Research*, 1–13.

Dearing, R. (1997). *Higher education in the learning society.* Retrieved from The National Committee of Enquiry into Higher Education: http://www.educationengland.org.uk/documents/dearing1997/dearing1997.html

Debnath, P. (2019). A peer into the philosophy of Dr Sarvepalli Radhakrishnan relating to education. *Pramana Research Journal*, 1235–1241.

Delener, N. (2013). Leadership excellence in higher education: Present and future. *Journal of Contemporary Issues in Business and Government*, 1–19.

Ebel, R. A. (2009). *Intellectual development models for undergraduate education.* Center for Teaching Excellence, United States Military Academy.

Faust, D. G. (2009, Septmebeer 1). The university's crisis of purpose. *The New York Times.* Retrieved from https://www.nytimes.com/2009/09/06/books/review/Faust-t.html

Fernandes Bella, R. L., Ferraz, F. T., & Bezerra, M. J. (2018). Workplace Spirituality: Sustainable Work Experience from a Human Factors Perspective. *Sustainability*, 1887. https://doi.org/10.3390/su10061887

Fox, C. R. (2016). A liberal education for the 21st century: Some reflections on general education. *Currents in Teaching & Learning*, 5–17.

Government of India, National Education Policy 2020. (2020). Retrieved from https://www.education.gov.in/sites/upload_files/mhrd/files/NEP_Final_English_0.pdf

Hammond, L.-D., Flook, L., Cook-Harvey, C., Barren, B., & Osher, D. (2019). Implications for educational practice of the science of learning and development. *Applied Developmental Science*, 97–140.

Hettich, P. (1998). *Cognitive development during and after college: Implications for access education.* Retrieved from https://files.eric.ed.gov/fulltext/ED424537.pdf

Houghton, A. M., & Anderson, J. (2017). Embedding mental wellbeing in the curriculum: Maximising success in higher education. *Higher Education Academy* (p. 44). https://s3.eu-west-2.amazonaws.com

Johnes, J. (2018). University Rankings: What do they really show? *Scientometrics*, 586–606. https://doi.org/10.1007/s11192-018-2666-1

Kaspura, A. (2019). *The engineering profession: A statistical overview.* Engineers Australia.

Keller, M. M., Hoy, A. W., Goetz, T., & Frenzel, A. C. (2015). Teacher enthusiasm: Reviewing and redefining a complex construct. *Educational Psychology Review*, 743–769.
King, P. M., & Kitchener, K. S. (2004). Reflective judgment: Theory and research on the development of epistemic assumptions through adulthood. *Educational Psychologist*, 5–18.
Kuriakose, F. (2018). Possibilities of a professor: An academic in the twenty first century. *Higher Education for the Future.*
Lambert, L. (1998). *Building leadership capacity in schools.* Association for Supervision & Curriculum.
Mahipalan, M. (2019). workplace spirituality and subjective happiness among high school teachers: Gratitude as a moderator. *Explore*, 107–114.
Mahler, D., Großschedl, J., & Harms, U. (2018). Does motivation matter?—The relationship between teachers' self-efficacy and enthusiasm and students' performance. *Plos One.* https://doi.org/10.1371/journal.pone.0207252
Margaretha, M., Saragih, S., Zaniarti, S., & Parayow, B. (2021). Workplace spirituality, employee engagement, and professional commitment: A study of lecturers from Indonesian universities. *Problems and Perspectives in Management*, 346–356.
Mental Health and Development. (2024). Retrieved from UN, Department of Economic and Social Affairs Disability : https://www.un.org/development/desa/disabilities/issues/mental-health-and-development.html
Merriam-Webster. (n.d.). *Accountability.* Retrieved from Merriam-Webster.com Dictionary: https://www.merriam-webster.com/dictionary/accountability
Moore, T., & Morton, J. (2015). The myth of job readiness? Written communication, employability, and the 'skills gap' in higher education. *Studies in Higher Education*, 591–601. https://doi.org/10.1080/03075079.2015.1067602
Moorkerji, R. K. (1941). Practical aspects of education in ancient India. In *Proceedings of the Indian History Congress* (pp. 127–34).
Mulcahy, D. G. (2009). What should it mean to have a liberal education in the 21st century? *Curriculum Inquiry*, 465–486.
Nathan, S. (2022). *The heart of work.* HarperBuisness.
Osborne, J., & Pimentel, D. (2022). Science, misinformation, and the role of education. *Science*, 1–9.
O'Shea, S., & Delahunty, J. (2018). Getting through the day and still having a smile on my face! How do students define success in the university learning environment? *Higher Education Research and Development.* https://doi.org/10.1080/07294360.2018.1463973
Patrick, B. C., Hisley, J., & Kempler, T. (2000). "What's everybody so excited about?": The effects of teacher enthusiasm on student intrinsic motivation and vitality. *The Journal of Experimental Education*, 217–236.
Patwardhan, B., & Mohanan, K. (2022). *Re-imagining assessment and accreditation in higher education in India.* National Assessment and Accreditation Council.
Pinkney, A. M. (2014). Looking West to India: Asian education, intra-Asian renaissance, and the Nalanda revival. *Modern Asian Studies*, 111–149.
PTI. (2023, July 26). *98 students died by suicide in higher education institutions in past five years: Govt.* Retrieved from Deccan Herald: https://www.deccanherald.com/india/98-students-died-by-suicide-in-higher-education-institutions-in-past-five-years-govt-1241013.html
Russell, J. F. (2008). Enthusiastic educational leadership. *Florida Journal of Educational Administration & Policy*, 79–96.
Scott, C. L. (2015). *The futures of learning 2-what kind of learning for the 21st century?* (ERF Working Papers, 1–23).
Selvan, K. T. (2023, December). Why we teach-Part I. *IEEE Antennas and Propagation Magazine*, pp. 48–51.
Selvan, K. T., & Warnick, K. F. (2023). Opinion: engineering program accreditation—Where have we been and where should we go? *Advances in Engineering Education*, 2–11.

Selvan, K. T., Stickel, M., Espinosa, H., Jha, S. K., Serre, A. J., & Gratchev, I. (2025). What do undergraduate students expect from their engineering education experience? Perspectives from an international survey, *European Journal of Engineering Education* [Submitted].

Shrikant, A. (2023, December 5). Youth suicide rates rose 62% from 2007 to 2021: 'People feel hopeless,' one recent grad says. *CNBC*. https://www.cnbc.com/2023/12/05/youth-suicide-rates-rose-62percent-from-2007-to-2021.html

Sofaer, S. (1999, December). Qualitative methods: What are they and why use them? *HSR Health Service Research*, 1101–1118.

Strauss, L. (2011). What is liberal education? In P. K. Robin Barrow (Ed.), *Academic ethics*. Ashgate Publishing.

thirukural.net. (2021). *The possession of knowledge*. Retrieved from Kural: https://www.thirukkural.net/en/kural/adhigaram-043.html

Vass, S. (2017, December 5). *Internetization: A new word for our global economy*. Retrieved from The Conversation: https://theconversation.com/internetization-a-new-word-for-our-global-economy-88013

Venkatachalam, R. (2015). *Thirukkural—translation—explanation:A life skills coaching approach*. Partridge Publishing India. Retrieved from thirukkural.io: https://thirukkural.io/43/arivudaimai/

Vidal, J., & Ferreira, C. (2020). Universities under pressure: the impact of international university rankings. *Journal of New Approaches in Educational Research*, 181–193.

WHO. (2023, August 28). *Suicide*. Retrieved from World Health Organization : https://www.who.int/news-room/fact-sheets/detail/suicide#:~:text=Suicide%20is%20the%20fourth%20leading,common%20methods%20of%20suicide%20globally

William Perry's Scheme of Intellectual and Ethical Development. (2003, January 29). Retrieved from James Madison University: https://www.jmu.edu/geology/_files/willperry21.pdf

Williamson, B., Bayne, S., & Shay, S. (2020). The datafication of teaching in higher education: Critical issues and perspectives. *Teaching in Higher Education*, 351–365.

Zalec, B. (2014). Trust, accountability, and higher education. *Synthesis Philosophica, 28*, 65–81.

MIX
Papier aus verantwortungsvollen Quellen
Paper from responsible sources
FSC® C105338

FSC
www.fsc.org

If you have any concerns about our products,
you can contact us on
ProductSafety@springernature.com

In case Publisher is established outside the EU,
the EU authorized representative is:
Springer Nature Customer Service Center GmbH
Europaplatz 3, 69115 Heidelberg, Germany

Printed by Libri Plureos GmbH
in Hamburg, Germany